ARTILLERY
THROUGH THE AGES

ARTILLERY
THROUGH THE AGES

A Short Illustrated History of Cannon,

Emphasizing Types Used in America

by

ALBERT MANUCY

Historian
Southeastern National Monuments

Drawings by Author

Technical Review by Harold L. Peterson

National Park Service Interpretive Series
History No. 3

University Press of the Pacific
Honolulu, Hawaii

Artillery Through the Ages:
A Short Illustred History of Cannon, Emphasizing
Types Used in America

by
Albert Manucy

ISBN: 0-89875-446-1

Reprinted from the 1962 edition

University Press of the Pacific
Honolulu, Hawaii
http://www.universitypressofthepacific.com

Contents

"PIERRIERS VULGARLY CALLED PATTEREROS,"
from Francis Grose, Military Antiquities, 1796.

The Era of Artillery

Looking at an old-time cannon, most people are sure of just one thing: the shot came out of the front end. For that reason these pages are written; people are curious about the fascinating weapon that so prodigiously and powerfully lengthened the warrior's arm. And theirs is a justifiable curiosity, because the gunner and his "art" played a significant role in our history.

THE ANCIENT ENGINES OF WAR

To compare a Roman catapult with a modern trench mortar seems absurd. Yet the only basic difference is the kind of energy that sends the projectile on its way.

In the dawn of history, war engines were performing the function of artillery (which may be loosely defined as a means of hurling missiles too heavy to be thrown by hand), and with these crude weapons the basic principles of artillery were laid down. The Scriptures record the use of ingenious machines on the walls of Jerusalem eight centuries B. C.—machines that were probably predecessors of the catapult and ballista, getting power from twisted ropes made of hair, hide or sinew. The ballista had horizontal arms like a bow. The arms were set in rope; a cord, fastened to the arms like a bowstring, fired arrows, darts, and stones. Like a modern field gun, the ballista shot low and directly toward the enemy.

The catapult was the howitzer, or mortar, of its day and could throw a hundred-pound stone 600 yards in a high arc to strike the enemy behind his wall or batter down his defenses. "In the middle of the ropes a wooden arm rises like a chariot pole," wrote the historian Marcellinus. "At the top of the arm hangs a sling. When battle is commenced, a round stone is set in the sling. Four soldiers on each side of the engine wind the arm down until it is almost level with the ground. When the arm is set free, it springs up and hurls the stone forth from its sling." In early times the weapon was called a "scorpion," for like this dreaded insect it bore its "sting" erect.

1

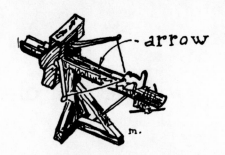

FIGURE 1—BALLISTA. Caesar covered his landing in Britain with fire from catapults and ballistas.

The trebuchet was another war machine used extensively during the Middle Ages. Essentially, it was a seesaw. Weights on the short arm swung the long throwing arm.

These weapons could be used with telling effect, as the Romans learned from Archimedes in the siege of Syracuse (214-212 B.C.). As Plutarch relates, "Archimedes soon began to play his engines upon the Romans and their ships, and shot stones of such an enormous size and with so incredible a noise and velocity that nothing could stand before them. At length

FIGURE 2—CATAPULT.

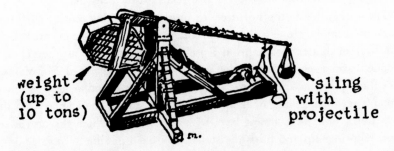

FIGURE 3—TREBUCHET. A heavy trebuchet could throw a 300-pound stone 300 yards.

the Romans were so terrified that, if they saw but a rope or a beam projecting over the walls of Syracuse, they cried out that Archimedes was leveling some machine at them, and turned their backs and fled."

Long after the introduction of gunpowder, the old engines of war continued in use. Often they were side by side with cannon.

GUNPOWDER COMES TO EUROPE

Chinese "thunder of the earth" (an effect produced by filling a large bombshell with a gunpowder mixture) sounded faint reverberations amongst the philosophers of the western world as early as A.D. 300. Though the Chinese were first instructed in the scientific casting of cannon by missionaries during the 1600's, crude cannon seem to have existed in China during the twelfth century and even earlier.

In Europe, a ninth century Latin manuscript contains a formula for gunpowder. But the first show of firearms in western Europe may have been by the Moors, at Saragossa, in A.D. 1118. In later years the Spaniards turned the new weapon against their Moorish enemies at the siege of Cordova (1280) and the capture of Gibraltar (1306).

It therefore follows that the Arabian *madfaa,* which in turn had doubtless descended from an eastern predecessor, was the original cannon brought to western civilization. This strange weapon seems to have been a small, mortar-like instrument of wood. Like an egg in an egg cup, the ball rested on the muzzle end until firing of the charge tossed it in the general direction of the enemy. Another primitive cannon, with narrow neck and flared mouth, fired an iron dart. The shaft of the dart was wrapped with leather to fit tightly into the neck of the piece. A red-hot bar thrust through a vent ignited the charge. The range was about 700 yards. The bottle shape of the weapon perhaps suggested the name *pot de fer* (iron jug) given early cannon, and in the course of evolution the narrow neck probably enlarged until the bottle became a straight tube.

During the Hundred Years' War (1339-1453) cannon came into general use. Those early pieces were very small, made of iron or cast bronze, and fired lead or iron balls. They were laid directly on the ground, with muzzles elevated by mounding up the earth. Being cumbrous and inefficient, they played little part in battle, but were quite useful in a siege.

THE BOMBARDS

By the middle 1400's the little popguns that tossed one- or two-pound pellets had grown into enormous bombards. Dulle Griete, the giant bombard of Ghent, had a 25-inch caliber and fired a 700-pound granite ball. It was built in 1382. Edinburgh Castle's famous Mons Meg threw a 19½-inch iron ball some 1,400 yards (a mile is 1,760 yards), or a stone ball twice that far.

The Scottish kings used Meg between 1455 and 1513 to reduce the castles of rebellious nobles. A baron's castle was easily knocked to pieces by the prince who owned, or could borrow, a few pieces of heavy ordnance. The towering walls of the old-time strongholds slowly gave way to the earthwork-protected Renaissance fortification, which is typified in the United States by Castillo de San Marcos, in Castillo de San Marcos National Monument, St. Augustine, Fla.

Some of the most formidable bombards were those of the Turks, who used exceptionally large cast-bronze guns at the siege of Constantinople in 1453. One of these monsters weighed 19 tons and hurled a 600-pound stone seven times a day. It took some 60 oxen and 200 men to move this

FIGURE 4—EARLY SMALL BOMBARD (1330). It was made of wrought-iron bars, bound with hoops.

piece, and the difficulty of transporting such heavy ordnance greatly reduced its usefulness. The largest caliber gun on record is the Great Mortar of Moscow. Built about 1525, it had a bore of 36 inches, was 18 feet long, and fired a stone projectile weighing a ton. But by this time the big guns were obsolete, although some of the old Turkish ordnance survived the centuries to defend Constantinople against a British squadron in 1807. In that defense a great stone cut the mainmast of the British flagship, and another crushed through the English ranks to kill or wound 60 men.

The ponderosity of the large bombards held them to level land, where they were laid on rugged mounts of the heaviest wood, anchored by stakes driven into the ground. A gunner would try to put his bombard 100 yards from the wall he wanted to batter down. One would surmise that the gunner, being so close to a castle wall manned by expert Genoese crossbowmen, was in a precarious position. He was; but earthworks or a massive wooden shield arranged like a seesaw over his gun gave him fair protection. Lowering the front end of the shield made a barricade behind which he could charge his muzzle loader (see fig. 49).

In those days, and for many decades thereafter, neither gun crews nor transport were permanent. They had to be hired as they were needed. Master gunners were usually civilian "artists," not professional soldiers, and many of them had cannon built for rental to customers. Artillerists obtained the right to captured metals such as tools and town bells, and this loot would be cast into guns or ransomed for cash. The making of

4

guns and gunpowder, the loading of bombs, and even the serving of cannon were jealously guarded trade secrets. Gunnery was a closed corporation, and the gunner himself a guildsman. The public looked upon him as something of a sorcerer in league with the devil, and a captured artilleryman was apt to be tortured and mutilated. At one time the Pope saw fit to excommunicate all gunners. Also since these specialists kept to themselves and did not drink or plunder, their behavior was ample proof to the good soldier of the old days that artillerists were hardly human.

SIXTEENTH CENTURY CANNON

After 1470 the art of casting greatly improved in Europe. Lighter cannon began to replace the bombards. Throughout the 1500's improvement was mainly toward lightening the enormous weights of guns and projectiles, as well as finding better ways to move the artillery. Thus, by 1556 Emperor Ferdinand was able to march against the Turks with 57 heavy and 127 light pieces of ordnance.

At the beginning of the 1400's cast-iron balls had made an appearance. The greater efficiency of the iron ball, together with an improvement in gunpowder, further encouraged the building of smaller and stronger guns. Before 1500 the siege gun had been the predominant piece. Now forged-iron cannon for field, garrison, and naval service—and later, cast-iron pieces—were steadily developed along with cast-bronze guns, some of which were beautifully ornamented with Renaissance workmanship. The casting of trunnions on the gun made elevation and transportation easier, and the cumbrous beds of the early days gave way to crude artillery carriages with trails and wheels. The French invented the limber and about 1550 took a sizable forward step by standardizing the calibers of their artillery.

Meanwhile, the first cannon had come to the New World with Columbus. As the *Pinta's* lookout sighted land on the early morn of October 12, 1492, the firing of a lombard carried the news over the moonlit waters to the flagship *Santa Maria*. Within the next century, not only the galleons, but numerous fortifications on the Spanish Main were armed with guns, thundering at the freebooters who disputed Spain's ownership of American treasure. Sometimes the adventurers seized cannon as prizes, as did Drake in 1586 when he made off with 14 bronze guns from St. Augustine's little wooden fort of San Juan de Pinos. Drake's loot no doubt included the ordnance of a 1578 list, which gives a fair idea of the armament for an important frontier fortification: three reinforced cannon, three demiculverins, two sakers (one broken), a demisaker and a falcon, all properly mounted on elevated platforms in the fort to cover every approach. Most of them were highly ornamented pieces founded between 1546 and 1555. The reinforced cannon, for instance, which seem to have been cast from the same mold, each bore the figure of a savage hefting a club in one hand and grasping a coin in the other. On a demiculverin, a bronze mermaid held a turtle, and the other guns were decorated with arms, escutcheons, the founder's name, and so on.

In the English colonies during the sixteenth and seventeenth centuries, lighter pieces seem to have been the more prevalent; there is no record of any "cannon." (In those days, "cannon" were a special class.) Culverins are mentioned occasionally and demiculverins rather frequently, but most common were the falconets, falcons, minions, and sakers. At Fort Raleigh, Jamestown, Plymouth, and some other settlements the breech-loading half-pounder perrier or "Patterero" mounted on a swivel was also in use. (See frontispiece.)

It was during the sixteenth century that the science of ballistics had its beginning. In 1537, Niccolo Tartaglia published the first scientific treatise on gunnery. Principles of construction were tried and sometimes abandoned, only to reappear for successful application in later centuries. Breech-loading guns, for instance, had already been invented. They were unsatisfactory because the breech could not be sealed against escape of the powder gases, and the crude, chambered breechblocks, jammed against the bore with a wedge, often cracked under the shock of firing. Neither is spiral rifling new. It appeared in a few guns during the 1500's.

Mobile artillery came on the field with the cart guns of John Zizka during the Hussite Wars of Bohemia (1419-24). Using light guns, hauled by the best of horses instead of the usual oxen, the French further improved field artillery, and maneuverable French guns proved to be an excellent means for breaking up heavy masses of pikemen in the Italian campaigns of the early 1500's. The Germans under Maximilian I, however, took the armament leadership away from the French with guns that ranged 1,500 yards and with men who had earned the reputation of being the best gunners in Europe.

Then about 1525 the famous Spanish Square of heavily armed pikemen and musketeers began to dominate the battlefield. In the face of musketry, field artillery declined. Although artillery had achieved some mobility, carriages were still cumbrous. To move a heavy English cannon, even over good ground, it took 23 horses; a culverin needed nine beasts. Ammunition—mainly cast-iron round shot, the bomb (an iron shell filled with gunpowder), canister (a can filled with small projectiles), and grape shot (a cluster of iron balls)—was carried the primitive way, in wheelbarrows and carts or on a man's back. The gunner's pace was the measure of field artillery's speed: the gunner *walked* beside his gun! Furthermore, some of these experts were getting along in years. During Elizabeth's reign several of the gunners at the Tower of London were over 90 years old.

Lacking mobility, guns were captured and recaptured with every changing sweep of the battle; so for the artillerist generally, this was a difficult period. The actual commander of artillery was usually a soldier; but transport and drivers were still hired, and the drivers naturally had a layman's attitude toward battle. Even the gunners, those civilian artists who owed

6

no special duty to the prince, were concerned mainly over the safety of their pieces—and their hides, since artillerists who stuck with their guns were apt to be picked off by an enemy musketeer. Fusilier companies were organized as artillery guards, but their job was as much to keep the gun crew from running away as to protect them from the enemy.

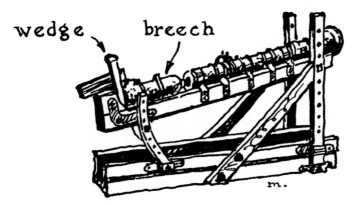

FIGURE 5—FIFTEENTH-CENTURY BREECHLOADER.

So, during 400 years, cannon had changed from the little vases, valuable chiefly for making noise, into the largest caliber weapons ever built, and then from the bombards into smaller, more powerful cannon. The gun of 1600 could throw a shot almost as far as the gun of 1850; not in fire power, but in mobility, organization, and tactics was artillery undeveloped. Because artillery lacked these things, the pike and musket were supreme on the battlefield.

THE SEVENTEENTH CENTURY AND GUSTAVUS ADOLPHUS

Under the Swedish warrior Gustavus Adolphus, artillery began to take its true position on the field of battle. Gustavus saw the need for mobility, so he divorced anything heavier than a 12-pounder from his field artillery. His famous "leatheren" gun was so light that it could be drawn and served by two men. This gun was a wrought-copper tube screwed into a chambered brass breech, bound with four iron hoops. The copper tube was covered with layers of mastic, wrapped firmly with cords, then coated with an equalizing layer of plaster. A cover of leather, boiled and varnished, completed the gun. Naturally, the piece could withstand only a small charge, but it was highly mobile.

Gustavus abandoned the leather gun, however, in favor of a cast-iron 4-pounder and a 9-pounder demiculverin produced by his bright young artillery chief, Lennart Torstensson. The demiculverin was classed as the "feildpeece" *par excellence*, while the 4-pounder was so light (about 500 pounds) that two horses could pull it in the field.

These pieces could be served by three men. Combining the powder charge and projectile into a single cartridge did away with the old method of ladling the powder into the gun and increased the rapidity of fire. Whereas in the past one cannon for each thousand infantrymen had been standard, Gustavus brought the ratio up to six cannon, and attached a pair of light pieces to each regiment as "battalion guns." At the same time he knew the value of fire concentration, and he frequently massed

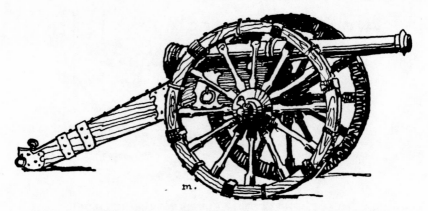

FIGURE 6—LIGHT ARTILLERY OF GUSTAVUS ADOLPHUS (1630).

guns in strong batteries. His plans called for smashing hostile infantry formations with artillery fire, while neutralizing the ponderous, immobile enemy guns with a whirlwind cavalry charge. The ideas were sound. Gustavus smashed the Spanish Squares at Breitenfeld in 1631.

Following the Swedish lead, all nations modified their artillery. Leadership fell alternately to the Germans, the French, and the Austrians. The mystery of artillery began to disappear, and gunners became professional soldiers. Bronze came to be the favorite gunmetal.

Louis XIV of France seems to have been the first to give permanent organization to the artillery. He raised a regiment of artillerymen in 1671 and established schools of instruction. The "standing army" principle that began about 1500 was by now in general use, and small armies of highly trained professional soldiers formed a class distinct from the rest of the population. As artillery became an organized arm of the military, expensive personnel and equipment had to be maintained even in peacetime. Still, some necessary changes were slow in coming. French artillery officers did not receive military rank until 1732, and in some countries drivers were still civilians in the 1790's. In 1716, Britain had organized artillery into two permanent companies, comprising the Royal Regiment of Artillery. Yet as late as the American Revolution there was a dispute about whether a general officer whose service had been in the Royal Artillery was entitled to command troops of all arms. There was no such question in England of the previous century: the artillery general was a personage having "always a part of the charge, and when the chief generall is ab sent, he is to command all the army."

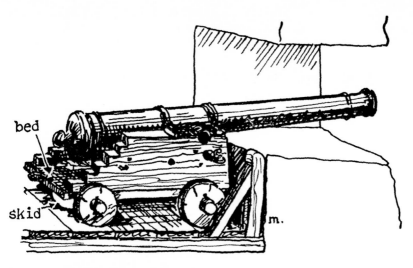

bed

skid

m.

FIGURE 7—FRENCH GARRISON GUN (1650-1700). The gun is on a sloping wooden platform at the embrasure. Note the heavy bed on which the cheeks of the carriage rest and the built-in skid under the center of the rear axletree.

THE EIGHTEENTH CENTURY

During the early 1700's cannon were used to protect an army's deployment and to prepare for the advance of the troops by firing upon enemy formations. There was a tendency to regard heavy batteries, properly protected by field works or permanent fortifications, as the natural role for artillery. But if artillery was seldom decisive in battle, it nevertheless waxed more important through improved organization, training, and discipline. In the previous century, calibers had been reduced in number and more or less standardized; now, there were notable scientific and technical improvements. The English scientist Benjamin Robins wedded theory to practice; his *New Principles of Gunnery* (1742) did much to bring about a more scientific attitude toward ballistics. One result of Robins' research was the introduction, in 1779, of carronades, those short, light pieces so useful in the confines of a ship's gun deck. Carronades usually ranged in caliber from 6- to 68-pounders.

In North America, cannon were generally too cumbrous for Indian fighting. But from the time (1565) the French, in Florida, loosed the first bolt at the rival fleet of the Spaniard Menéndez, cannon were used on land and sea during intercolonial strife, or against corsairs. Over the vast distances of early America, transport of heavy guns was necessarily by water. Without ships, the guns were inexorably walled in by the forest. So it was when the Carolinian Moore besieged St. Augustine in 1702. When his ships burned, Moore had to leave his guns to the Spaniards.

One of the first appearances of organized American field artillery on the battlefield was in the Northeast, where France's Louisburg fell to British and Colonial forces in 1745. Serving with the British Royal Artillery was the Ancient and Honorable Artillery Company of Boston, which had originated in 1637. English field artillery of the day had "brigades" of four to six cannon, and each piece was supplied with 100 rounds of solid shot and 30 rounds of grape. John Müller's *Treatise on Artillery*, the standard English authority, was republished in Philadelphia (1779), and British artillery was naturally a model for the arm in America.

FIGURE 8—AMERICAN 6-POUNDER FIELDPIECE (c. 1775).

At the outbreak of the War of Independence, American artillery was an accumulation of guns, mortars, and howitzers of every sort and some 13 different calibers. Since the source of importation was cut off, the undeveloped casting industries of the Colonies undertook cannon founding, and by 1775 the foundries of Philadelphia were casting both bronze and iron guns. A number of bronze French guns were brought in later. The mobile guns of Washington's army ranged from 3- to 24-pounders, with 5½- and 8-inch howitzers. They were usually bronze. A few iron siege guns of 18-, 24-, and 32-pounder caliber were on hand. The guns used round shot, grape, and case shot; mortars and howitzers fired bombs and carcasses. "Side boxes" on each side of the carriage held 21 rounds of ammunition and were taken off when the piece was brought into battery. Horses or oxen, with hired civilian drivers, formed the transport. On the battlefield the cannoneers manned drag ropes to maneuver the guns into position.

Sometimes, as at Guilford Courthouse, the ever-present forest diminished the effectiveness of artillery, but nevertheless the arm was often put to good use. The skill of the American gunners at Yorktown contributed no little toward the speedy advance of the siege trenches. Yorktown battlefield today has many examples of Revolutionary War cannon, including some fine ship guns recovered from British vessels sunk during the siege of 1781.

In Europe, meanwhile, Frederick the Great of Prussia learned how to use cannon in the campaigns of the Seven Years' War (1756-63). The education was forced upon him as gradual destruction of his veteran in-

10

fantry made him lean more heavily on artillery. To keep pace with cavalry movements, he developed a horse artillery that moved rapidly along with the cavalry. His field artillery had only light guns and howitzers. With these improvements he could establish small batteries at important points in the battle line, open the fight, and protect the deployment of his columns with light guns. What was equally significant, he could change the position of his batteries according to the course of the action.

Frederick sent his 3- and 6-pounders ahead of the infantry. Gunners dismounted 500 paces from the enemy and advanced on foot, pushing their guns ahead of them, firing incessantly and using grape shot during the latter part of their advance. Up to closest range they went, until the infantry caught up, passed through the artillery line, and stormed the en- ‑my position. Remember that battle was pretty formal, with musketeers s anding or kneeling in ranks, often in full view of the enemy!

FIGURE 9—FRENCH 12-POUNDER FIELD GUN (c. 1780).

Perhaps the outstanding artilleryman of the 1700's was the Frenchman Jean Baptiste de Gribeauval, who brought home a number of ideas after serving with the capable Austrian artillery against Frederick. The great reform in French artillery began in 1765, although Gribeauval was not able to effect all of his changes until he became Inspector General of Artillery in 1776. He all but revolutionized French artillery, and vitally influenced other countries.

Gribeauval's artillery came into action at a gallop and smothered enemy batteries with an overpowering volume of fire. He created a distinct matériel for field, siege, garrison, and coast artillery. He reduced the length and weight of the pieces, as well as the charge and the windage (the difference between the diameters of shot and bore); he built carriages so that many parts were interchangeable, and made soldiers out of the drivers. For siege and garrison he adopted 12- and 16-pounder guns, an 8-inch howitzer and 8-, 10-, and 12-inch mortars. For coastal fortifications he used the traversing platform which, having rear wheels that ran upon a track, greatly simplified the training of a gun right or left upon a moving target (fig. 10). Gribeauval-type matériel was used with the greatest effect in the new tactics which Napoleon introduced.

Napoleon owed much of his success to masterly use of artillery. Under this captain there was no preparation for infantry advance by slowly disintegrating the hostile force with artillery fire. Rather, his artillerymen went up fast into closest range, and by actually annihilating a portion of the enemy line with case-shot fire, covered the assault so effectively that columns of cavalry and infantry reached the gap without striking a blow!

After Napoleon, the history of artillery largely becomes a record of its technical effectiveness, together with improvements or changes in putting well-established principles into action.

UNITED STATES GUNS OF THE EARLY 1800's

The United States adopted the Gribeauval system of artillery carriages in 1809, just about the time it was becoming obsolete (the French abandoned it in 1829). The change to this system, however, did not include adoption of the French gun calibers. Early in the century cast iron replaced bronze as a gunmetal, a move pushed by the growing United States iron industry; and not until 1836 was bronze readopted in this

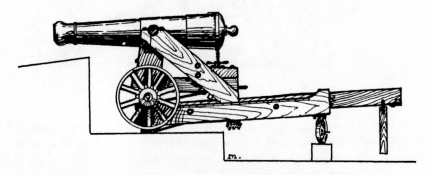

FIGURE 10—U. S. 32-POUNDER ON BARBETTE CARRIAGE (1860).

country for mobile cannon. In the meantime, U. S. Artillery in the War of 1812 did most of its fighting with iron 6-pounders. Fort McHenry, which is administered by the National Park Service as a national monument and historic shrine, has a few ordnance pieces of the period.

During the Mexican War, the artillery carried 6-and 12-pounder guns, the 12-pounder mountain howitzer (a light piece of 220 pounds which had been added for the Indian campaigns), a 12-pounder field howitzer (788 pounds), the 24- and 32-pounder howitzers, and 8- and 10-inch mortars. For siege, garrison, and seacoast there were pieces of 16 types, ranging from a 1-pounder to the giant 10-inch Columbiad of 7½ tons. In 1857, the United States adopted the 12-pounder Napoleon gun-howitzer, a bronze smoothbore designed by Napoleon III, and this muzzle-loader remained standard in the army until the 1880's.

The naval ironclads, which were usually armed with powerful 11- or 15-inch smoothbores, were a revolutionary development in mid-century. They were low-hulled, armored, steam vessels, with one or two revolving turrets. Although most cannonballs bounced from the armor, lack of speed made the "cheese box on a raft" vulnerable, and poor visibility through the turret slots was a serious handicap in battle.

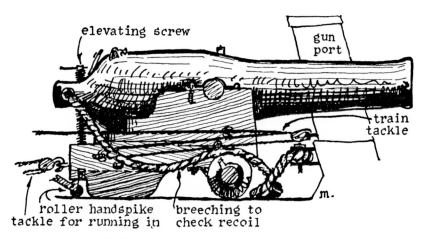

Figure 11—U. S. NAVY 9-INCH SHELL-GUN ON MARSILLY CARRIAGE (1866).

While 20-, 30-, and 60-pounder Parrott rifles soon made an appearance in the Federal Navy, along with Dahlgren's 12- and 20-pounder rifled howitzers, the Navy relied mainly upon its "shell-guns": the 9-, 10-, 11-, and 15-inch iron smoothbores. There were also 8-inch guns of 55 and 63 "hundredweight" (the contemporary naval nomenclature), and four sizes of 32-pounders ranging from 27 to 57 hundredweight. The heavier guns took more powder and got slightly longer ranges. Many naval guns of the period are characterized by a hole in the cascabel, through which the breeching tackle was run to check recoil. The Navy also had a 13-inch mortar, mounted aboard ship on a revolving circular platform. Landing parties were equipped with 12- or 24-pounder howitzers either on boat carriages (a flat bed something like a mortar bed) or on three-wheeled "field" carriages.

RIFLING

Rifling, by imparting a spin to the projectile as it travels along the spiral grooves in the bore, permits the use of a long projectile and ensures its flight point first, with great increase in accuracy. The longer projectile, being both heavier and more streamlined than round shot of the same caliber, also has a greater striking energy.

Though Benjamin Robins was probably the first to give sound reasons, the fact that rifling was helpful had been known a long time. A 1542 barrel at Woolwich has six fine spiral grooves in the bore. Straight grooving had been applied to small arms as early as 1480, and during the 1500's straight grooving of musket bores was extensively practiced. Probably, rifling evolved from the early observation of the feathers on an arrow—and from the practical results of cutting channels in a musket, originally to reduce fouling, then because it was found to improve accuracy of the shot. Rifled small-arm efficiency was clearly shown at Kings Mountain during the American Revolution.

In spite of earlier experiments, however, it was not until the 1840's that attempts to rifle cannon could be called successful. In 1846, Major Cavelli in Italy and Baron Wahrendorff in Germany independently produced rifled iron breech-loading cannon. The Cavelli gun had two spiral grooves into which fitted the ¼-inch projecting lugs of a long projectile (fig. 12a). Other attempts at what might be called rifling were Lancaster's elliptical-bore gun and the later development of a spiraling hexagonal-bore by Joseph Whitworth (fig. 12b). The English Whitworth was used by Confederate artillery. It was an efficient piece, though subject to easy fouling that made it dangerous.

Then, in 1855, England's Lord Armstrong designed a rifled breech-loader that included so many improvements as to be revolutionary. This gun was rifled with a large number of grooves and fired lead-coated projectiles. Much of its success, however, was due to the built-up construction: hoops were shrunk on over the tube, with the fibers of the metal running in the directions most suitable for strength. Several United States muzzle-loading rifles of built-up construction were produced about the same time as the Armstrong and included the Chambers (1849), the Treadwell (1855), and the well-known Parrott of 1861 (figs. 12e and 13).

The German Krupp rifle had an especially successful breech mechanism. It was not a built-up gun, but depended on superior crucible steel for its strength. Cast steel had been tried as a gunmetal during the sixteenth and seventeenth centuries, but metallurgical knowledge of the early days could not produce sound castings. Steel was also used in other mid-nineteenth century rifles, such as the United States Wiard gun and the British Blakely, with its swollen, cast-iron breech hoop. Fort Pulaski National Monument, near Savannah, Ga., has a fine example of a 24-pounder Blakely used by the Confederates in the 1862 defense of the fort.

The United States began intensive experimentation with rifled cannon late in the 1850's, and a few rifled pieces were made by the South Boston Iron Foundry and also by the West Point Foundry at Cold Spring, N. Y. The first appearance of rifles in any quantity, however, was near the outset of the 1861 hostilities, when the Federal artillery was equipped with 300 wrought-iron 3-inch guns (fig. 14e). This "12-pounder," which fired a 10-pound projectile, was made by wrapping sheets of boiler iron around

14

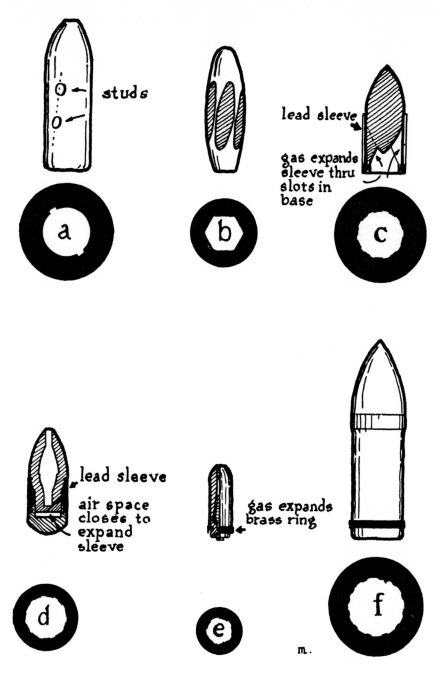

studs

lead sleeve

gas expands sleeve thru slots in base

lead sleeve

air space closes to expand sleeve

gas expands brass ring

m.

FIGURE 12—DEVELOPMENT OF RIFLE PROJECTILES (1840-1900). a—Cavelli type. b—Whitworth. c—James. d—Hotchkiss. e—Parrott. f—Copper rotating band type. (Not to scale.)

15

a mandrel. The cylinder thus formed was heated and passed through the rolls for welding, then cooled, bored, turned, and rifled. It remained in service until about 1900. Another rifle giving good results was the cast-iron 4½-inch siege gun. This piece was cast solid, then bored, turned, and rifled. Uncertainty of strength, a characteristic of cast iron, caused its later abandonment.

FIGURE 13—PARROTT 10-POUNDER RIFLE (1864).

The United States rifle that was most effective in siege work was the invention of Robert P. Parrott. His cast-iron guns (fig. 13), many of which are seen today in the battlefield parks, are easily recognized by the heavy wrought-iron jacket reinforcing the breech. The jacket was made by coiling a bar over the mandrel in a spiral, then hammering the coils into a welded cylinder. The cylinder was bored and shrunk on the gun. Parrotts were founded in 10-, 20-, 30-, 60-, 100-, 200-, and 300-pounder calibers, one foundry making 1,700 of them during the Civil War.

All nations, of course, had large stocks of smoothbores on hand, and various methods were devised to make rifles out of them. The U. S. Ordnance Board, for instance, believed the conversion simply involved cutting grooves in the bore, right at the forts or arsenals where the guns were. In 1860, half of the United States artillery was scheduled for conversion. As a result, a number of old smoothbores were rebored to fire rifle projectiles of the various patents which preceded the modern copper rotating band (fig. 12c, d, f). Under the James patent (fig. 12c) the weight of metal thrown by a cannon was virtually doubled; converted 24-, 32- and 42-pounders fired elongated shot classed respectively as 48-, 64-, and 84-pound projectiles. After the siege of Fort Pulaski, Federal Gen. Q. A. Gillmore praised the 84-pounder and declared "no better piece for breaching can be desired," but experience soon proved the heavier projectiles caused increased pressures which converted guns could not withstand for long.

The early United States rifles had a muzzle velocity about the same as the smoothbore, but whereas the round shot of the smoothbore lost speed so rapidly that at 2,000 yards its striking velocity was only about a third of the muzzle velocity, the more streamlined rifle projectile lost speed very slowly. But the rifle had to be served more carefully than the smoothbore.

16

Rifling grooves were cleaned with a moist sponge, and sometimes oiled with another sponge. Lead-coated projectiles like the James, which tended to foul the grooves of the piece, made it necessary to scrape the rifle grooves after every half dozen shots, although guns using brass-banded projectiles did not require the extra operation. With all muzzle-loading rifles, the projectile had to be pushed close home to the powder charge; otherwise, the blast would not fully expand its rotating band, the projectile would not take the grooves, and would "tumble" after leaving the gun, to the utter loss of range and accuracy. Incidentally, gunners had to "run out" (push the gun into firing position) both smoothbore and rifled muzzle-loaders carefully. A sudden stop might make the shot start forward as much as 2 feet.

When the U. S. Ordnance Board recommended the conversion to rifles, it also recommended that all large caliber iron guns be manufactured on the method perfected by Capt. T. J. Rodman, which involved casting the gun around a water-cooled core. The inner walls of the gun thus solidified first, were compressed by the contraction of the outer metal as it cooled down more slowly, and had much greater strength to resist explosion of the charge. The Rodman smoothbore, founded in 8-, 10-, 15-, and 20-inch calibers, was the best cast-iron ordnance of its time (fig. 14f). The 20-inch gun, produced in 1864, fired a 1,080-pound shot. The 15-incher was retained in service through the rest of the century, and these monsters are still to be seen at Fort McHenry National Monument and Historic Shrine or on the ramparts of Fort Jefferson, in the national monument of that name, in the Dry Tortugas Islands. In later years, a number of 10-inch Rodmans were converted into 8-inch rifles by enlarging the bore and inserting a grooved steel tube.

THE WAR BETWEEN THE STATES

At the opening of this civil conflict most of the matériel for both armies was of the same type—smoothbore. The various guns included weapons in the great masonry fortifications built on the long United States coast line since the 1820's—weapons such as the Columbiad, a heavy, long-chambered American muzzle-loader of iron, developed from its bronze forerunner of 1810. The Columbiad (fig. 14d) was made in 8-, 10-, and 12-inch calibers and could throw shot and shell well over 5,000 yards. "New" Columbiads came out of the foundries at the start of the 1860's, minus the powder chamber and with smoother lines. Behind the parapets or in fort gunrooms were 32- and 42-pounder iron seacoast guns (fig. 10); 24-pounder bronze howitzers lay in the bastions to flank the long reaches of the fort walls. There were 8-inch seacoast howitzers for heavier work. The largest caliber piece was the ponderous 13-inch seacoast mortar.

Siege and garrison cannon included 24-pounder and 8-inch bronze howitzers (fig. 14b), a 10-inch bronze mortar (fig. 14a), 12-, 18-, and 24-pounder iron guns (fig. 14c) and later the 4½-inch cast-iron rifle. With

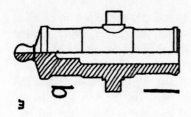
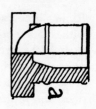

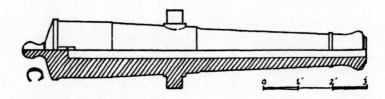

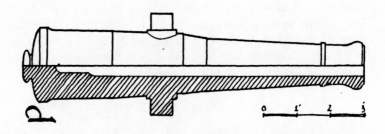

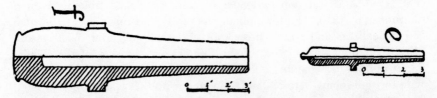

FIGURE 14—U. S. ARTILLERY TYPES (1861-1865). a—Siege mortar. b—
8-inch siege howitzer. c—24-pounder siege gun. d—8-inch Columbiad. e—
3-inch wrought-iron rifle. f—10-inch Rodman.

the exception of the new 3-inch wrought-iron rifle (fig. 14e), field artillery cannon were bronze: 6- and 12-pounder guns, the 12-pounder Na-

poleon gun-howitzer, 12-pounder mountain howitzer, 12-, 24-, and 32-pounder field howitzers, and the little Coehorn mortar (fig. 39). A machine gun invented by Dr. Richard J. Gatling became part of the artillery equipment during the war, but was not much used. Reminiscent of the ancient ribaudequin, a repeating cannon of several barrels, the Gatling gun could fire about 350 shots a minute from its 10 barrels, which were rotated and fired by turning a crank. In Europe it became more popular than the French mitrailleuse.

The smaller smoothbores were *effective* with case shot up to about 600 or 700 yards, and *maximum* range of field pieces went from something less than the 1,566-yard solid-shot trajectory of the Napoleon to about 2,600 yards (a mile and a half) for a 6-inch howitzer. At Chancellorsville, one of Stonewall Jackson's guns fired a shot which bounded down the center of a roadway and came to rest a mile away. The performance verified the drill-book tables. Maximum ranges of the larger pieces, however, ran all the way from the average 1,600 yards of an 18-pounder garrison gun to the well over 3-mile range of a 12-inch Columbiad firing a 180-pound shell at high elevation. A 13-inch seacoast mortar would lob a 200-pound shell 4,325 yards, or almost 2½ miles. The shell from an 8-inch howitzer carried 2,280 yards, but at such extreme ranges the guns could hardly be called accurate.

On the battlefield, Napoleon's artillery tactics were no longer practical. The infantry, armed with its own comparatively long-range firearm, was usually able to keep artillery beyond case-shot range, and cannon had to stand off at such long distances that their primitive ammunition was relatively ineffective. The result was that when attacking infantry moved in, the defending infantry and artillery were still fresh and unshaken, ready to pour a devastating point-blank fire into the assaulting lines. Thus, in spite of an intensive 2-hour bombardment by 138 Confederate guns at the crisis of Gettysburg, as the gray-clad troops advanced across the field to close range, double canister and concentrated infantry volleys cut them down in masses.

Field artillery smoothbores, under conditions prevailing during the war, generally gave better results than the smaller-caliber rifle. A 3-inch rifle, for instance, had twice the range of a Napoleon; but in the broken, heavily wooded country where so much of the fighting took place, the superior range of the rifle could not be used to full advantage. Neither was its relatively small and sometimes defective projectile as damaging to personnel as case or grape from a larger caliber smoothbore. At the first battle of Manassas (July 1861) more than half the 49 Federal cannon were rifled; but by 1863, even though many more rifles were in service, the majority of the pieces in the field were still the old reliable 6- and 12-pounder smoothbores.

It was in siege operations that the rifles forced a new era. As the smoke cleared after the historic bombardment of Fort Sumter in 1861, military men were already speculating on the possibilities of the newfangled weapon. A Confederate 12-pounder Blakely had pecked away at Sumter with amazing accuracy. But the first really effective use of the rifles in siege operations was at Fort Pulaski (1862). Using 10 rifles and 26 smoothbores, General Gillmore breached the 7½-foot-thick brick walls in little more than 24 hours. Yet his batteries were a mile away from the target! The heavier rifles were converted smoothbores, firing 48-, 64-, and 84-pound James projectiles that drove into the fort wall from 19 to 26 inches at each fair shot. The smoothbore Columbiads could penetrate only 13 inches, while from this range the ponderous mortars could hardly hit the fort. A year later, Gillmore used 100-, 200-, and 300-pounder Parrott rifles against Fort Sumter. The big guns, firing from positions some 2 miles away and far beyond the range of the fort guns, reduced Sumter to a smoking mass of rubble.

The range and accuracy of the rifles startled the world. A 30-pounder (4.2-inch) Parrott had an amazing carry of 8,453 yards with 80-pound hollow shot; the notorious "Swamp Angel" that fired on Charleston in 1863 was a 200-pounder Parrott mounted in the marsh 7,000 yards from the city. But strangely enough, neither rifles nor smoothbores could destroy earthworks. As was proven several times during the war, the defenders of a well-built earthwork were able to repair the trifling damage done by enemy fire almost as soon as there was a lull in the shooting. Learning this lesson, the determined Confederate defenders of Fort Sumter in 1863-64 refused to surrender, but under the most difficult conditions converted their ruined masonry into an earthwork almost impervious to further bombardment.

THE CHANGE INTO MODERN ARTILLERY

With Rodman's gun, the muzzle-loading smoothbore was at the apex of its development. Through the years great progress had been made in mobility, organization, and tactics. Now a new era was beginning, wherein artillery surpassed even the decisive role it had under Gustavus Adolphus and Napoleon. In spite of new infantry weapons that forced cannon ever farther to the rear, artillery was to become so deadly that its fire caused over 75 percent of the battlefield casualties in World War I.

Many of the vital changes took place during the latter years of the 1800's, as rifles replaced the smoothbores. Steel came into universal use for gun founding; breech and recoil mechanisms were perfected; smokeless powder and high explosives came into the picture. Hardly less important was the invention of more efficient sighting and laying mechanisms.

The changes did not come overnight. In Britain, after breechloaders had been in use almost a decade, the ordnance men went back to muzzle-loading rifles; faulty breech mechanisms caused too many accidents. Not until one of H.M.S. *Thunderer's* guns was inadvertently double-loaded did the English return to an improved breechloader.

The steel breechloaders of the Prussians, firing two rounds a minute with a percussion shell that broke into about 30 fragments, did much to defeat the French (1870-71). At Sedan, the greatest artillery battle fought prior to 1914, the Prussians used 600 guns to smother the French army. So thoroughly did these guns do their work that the Germans annihilated the enemy at the cost of only 5 percent casualties. It was a demonstration of using great masses of guns, bringing them quickly into action to destroy the hostile artillery, then thoroughly "softening up" enemy resistance in preparation for the infantry attack. While the technical progress of the Prussian artillery was considerable, it was offset in large degree by the counter-development of field entrenchment.

As the technique of forging large masses of steel improved, most nations adopted built-up (reinforcing hoops over a steel tube) or wire-wrapped steel construction for their cannon. With the advent of the metal cartridge case and smokeless powder, rapid-fire guns came into use. The new powder, first used in the Russo-Turkish War (1877-78), did away with the thick white curtain of smoke that plagued the gunner's aim, and thus opened the way for production of mechanisms to absorb recoil and return the gun automatically to firing position. Now, gunners did not have to lay the piece after every shot, and the rate of fire increased. Shields appeared on the gun—protection that would have been of little value in the days when gunners had to stand clear of a back-moving carriage.

During the early 1880's the United States began work on a modern system of seacoast armament. An 8-inch breech-loading rifle was built in 1883, and the disappearing carriage, giving more protection to both gun and crew, was adopted in 1886. Only a few of the weapons were installed by 1898; but fortunately the overwhelming naval superiority of the United States helped bring the War with Spain to a quick close.

During this war, United States forces were equipped with a number of British 2.95-inch mountain rifles, which, incidentally, served as late as

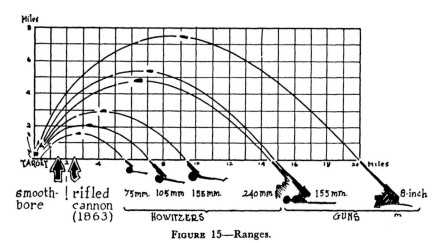

FIGURE 15—Ranges.

World War II in the pack artillery of the Philippine Scouts. Within the next few years the antiquated pieces such as the 3-inch wrought-iron rifle, the 4.2-inch Parrott siege gun, converted Rodmans, and the 15-inch Rodman smoothbore were finally pushed out of the picture by new steel guns. There were small-caliber rapid-fire guns of different types, a Hotchkiss 1.65-inch mountain rifle, and Hotchkiss and Gatling machine guns. The basic pieces in field artillery were 3.2- and 3.6-inch guns and a 3.6-inch mortar. Siege artillery included a 5-inch gun, 7-inch howitzers, and mortars. In seacoast batteries were 8-, 10-, 12-, 14-, and 16-inch guns and 12-inch mortars of the primary armament; intermediate rapid-fire guns of 4-, 4.72-, 5-, and 6-inch calibers; and 6- and 15-pounder rapid-fire guns in the secondary armament.

The Japanese showed the value of the French system of indirect laying (aiming at a target not visible to the gunner) during the Russo-Japanese War (1904-05). Meanwhile, the French 75-mm. gun of 1897, firing 6,000 yards, made all other field artillery cannon obsolete. In essence, artillery had assumed the modern form. The next changes were wrought by startling advances in motor transport, signal communications, chemical warfare, tanks, aviation, and mass production.

Gunpowder

Black powder was used in all firearms until smokeless and other type propellants were invented in the latter 1800's. "Black" powder (which was sometimes brown) is a mixture of about 75 parts saltpeter (potassium nitrate), 15 parts charcoal, and 10 parts sulphur by weight. It will explode because the mixture contains the necessary amount of oxygen for its own combustion. When it burns, it liberates smoky gases (mainly nitrogen and carbon dioxide) that occupy some 300 times as much space as the powder itself.

Early European powder "recipes" called for equal parts of the three ingredients, but gradually the amount of saltpeter was increased until Tartaglia reported the proportions to be 4-1-1. By the late 1700's "common war powder" was made 6-1-1, and not until the next century was the formula refined to the 75-15-10 composition in majority use when the newer propellants arrived on the scene.

As the name suggests, this explosive was originally in the form of powder or dust. The primitive formula burned slowly and gave low pressures—fortunate characteristics in view of the barrel-stave construction of the early cannon. About 1450, however, powder makers began to "corn" the powder. That is, they formed it into larger grains, with a resulting increase in the velocity of the shot. It was "corned" in fine grains for small arms and coarse for cannon.

Making corned powder was fairly simple. The three ingredients were pulverized and mixed, then compressed into cakes which were cut into "corns" or grains. Rolling the grains in a barrel polished off the corners; removing the dust essentially completed the manufacture. It has always been difficult, however, to make powder twice alike and keep it in condition, two factors which helped greatly to make gunnery an "art" in the old days. Powder residue in the gun was especially troublesome, and a disk-like tool (fig. 44) was designed to scrape the bore. Artillerymen at Castillo de San Marcos complained that the "heavy" powder from Mexico was especially bad, for after a gun was fired a few times, the bore was so fouled that cannonballs would no longer fit. The gunners called loudly for better grade powder from Spain itself.

23

How much powder to use in a gun has been a moot question through the centuries. According to the Spaniard Collado in 1592, the proper yardstick was the amount of metal in the gun. A legitimate culverin, for instance, was "rich" enough in metal to take as much powder as the ball weighed. Thus, a 30-pounder culverin would get 30 pounds of powder. Since a 60-pounder battering cannon, however, had in proportion a third less metal than the culverin, the charge must also be reduced by a third—to 40 pounds!

FIGURE 16—GUNPOWDER. Black powder (above) is a mechanical mixture; modern propellants are chemical compounds.

Other factors had to be taken into account, such as whether the powder was coarse- or fine-grained; and a short gun got less powder than a long one. The bore length of a legitimate culverin, said Collado, was 30 calibers (30 times the bore diameter), so its powder charge was the same as the weight of the ball. If the gunner came across a culverin only 24 calibers long, he must load this piece with only 24/30 of the ball's weight. Collado's *pasavolante* had a tremendous length of some 40 calibers and fired a 6- or 7-pound lead ball. Because it had plenty of metal "to resist, and the length to burn" the powder, it was charged with the full weight of the ball in fine powder, or three-fourths as much with cannon powder. The lightest charge seems to have been for the pedrero, which fired a stone ball. Its charge was a third of the stone's weight.

In later years, powder charges lessened for all guns. English velocity tables of the 1750's show that a 9-pounder charged with 2¼ pounds of powder might produce its ball at a rate of 1,052 feet per second. By almost tripling the charge, the velocity would increase about half. But the increase did not mean the shot hit the target 50 percent harder, for the higher the velocity, the greater was the air resistance; or as Müller phrased it: "a great quantity of Powder does not always produce a greater effect." Thus, from two-thirds the ball's weight, standard charges dropped to one-third or even a quarter; and by the 1800's they became even smaller. The United States manual of 1861 specified 6 to 8 pounds for a 24-pounder siege gun, depending on the range; a Columbiad firing 172-pound shot used only 20 pounds of powder. At Fort Sumter, Gillmore's rifles firing 80-pound shells used 10 pounds of powder. The rotating band on the rifle shell, of course, stopped the gases that had slipped by the loose-fitting cannonball.

Black powder was, and is, both dangerous and unstable. Not only is it sensitive to flame or spark, but it absorbs moisture from the air. In other words, it was no easy matter to "keep your powder dry." During the middle 1700's, Spaniards on a Florida river outpost kept powder in glass bottles; earlier soldiers, fleeing into the humid forest before Sir Francis Drake, carried powder in *peruleras*—stoppered, narrow-necked pitchers.

As for magazines, a dry magazine was just about as important as a shell-proof one. Charcoal and chloride of lime, hung in containers near the ceiling, were early used as dehydrators, and in the eighteenth century standard English practice was to build the floor 2 feet off the ground and lay stone chips or "dry sea coals" under the flooring. Side walls had air holes for ventilation, but screened to prevent the enemy from letting in some small animal with fire tied to his tail. Powder casks were laid on their sides and periodically rolled to a different position; "otherwise," explains a contemporary expert, "the salt petre, being the heaviest ingredient, will descend into the lower part of the barrel, and the powder above will lose much of its goodness."

FIGURE 17—SPANISH POWDER BUCKET (c. 1750).

In the dawn of artillery, loose powder was brought to the gun in a covered bucket, usually made of leather. The loader scooped up the proper amount with a ladle (fig. 44), and inserted it into the gun. He could, by using his experienced judgment, put in just enough powder to give him the range he wanted, much as our modern artillerymen sometimes use only a portion of their charge. After Gustavus Adolphus in the 1630's, however, powder bags came into wide use, although English gunners long preferred to ladle their powder. The powder bucket or "passing box" of course remained on the scene. It was usually large enough to hold a pair of cartridge bags.

The root of the word cartridge seems to be "carta," meaning paper. But paper was only one of many materials such as canvas, linen, parchment, flannel, the "woolen stuff" of the 1860's, and even wood. Until the advent of the silk cartridge, nothing was entirely satisfactory. The materials did not burn completely, and after several rounds it was mandatory to withdraw the unburnt bag ends with a wormer (fig. 44), else they accumulated to the point where they blocked the vent or "touch hole" by which the piece was fired. Parchment bags shriveled up and stuck in the vent, purpling many a good gunner's face.

25

When the powder bag came into use, the gunner had to prick the bag open so the priming fire from the vent could reach the charge. The operation was accomplished simply enough by plunging the gunner's pick into the vent far enough to pierce the bag. Then the vent was primed with loose powder from the gunner's flask. The vent prime, which was not much improved until the nineteenth century, was a trick learned from the fourteenth century Venetians. There were numerous tries for improvement, such as the powder-filled tin tube of the 1700's, the point of which pierced the powder bag. But for all of them, the slow match had to be used to start the fire train.

FIGURE 18—LINSTOCKS.

Before 1800, the slow match was in universal use for setting off the charge. The match was usually a 3-strand cotton rope, soaked in a solution of saltpeter and otherwise chemically treated with lead acetate and lye to burn very slowly—about 4 or 5 inches an hour. It was attached to a linstock (fig. 18), a forked stick long enough to keep the cannoneer out of the way of the recoil.

Chemistry advances, like the isolation of mercury fulminate in 1800, led to the invention of the percussion cap and other primers. On many a battleground you may have picked up a scrap of twisted wire—the loop of a friction primer. The device was a copper tube (fig. 19) filled with powder. The tube went into the vent of the cannon and buried its tip in the powder charge. Near the top of this tube was soldered a "spur"—a short tube containing a friction composition (antimony sulphide and potassium chlorate). Lying in the composition was the roughened end of a wire "slider." The other end of the slider was twisted into a loop for hooking to the gunner's lanyard. It was like striking a match: a smart pull on the lanyard, and the rough slider ignited the composition. Then the powder in the long tube began to burn and fired the charge in the cannon. Needless to say, it happened faster than we can tell it!

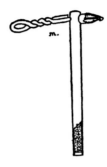

FIGURE 19—FRICTION PRIMER.

The percussion primer was even more simple: a "quill tube," filled with fine powder, fitted into the vent. A fulminate cap was glued to the top of the tube. A pull of the lanyard caused the hammer of the cannon to strike the cap (just like a little boy's cap pistol) and start the train of explosions.

Because the early methods of priming left the vent open when the cannon fired, the little hole tended to enlarge. Many cannon during the 1800's were made with two vents, side by side. When the first one wore out, it was plugged, and the second vent opened. Then, to stop this "erosion," the obturating (sealing) primer came into use. It was like the common friction primer, but screwed into and sealed the vent. Early electric primers, by the way, were no great departure from the friction primer; the wires fired a bit of guncotton, which in turn ignited the powder in the primer tube.

MODERN USE OF BLACK POWDER

Aside from gradual improvement in the formula, no great change in powder making came until 1860, when Gen. Thomas J. Rodman of the U. S. Ordnance Department began to tailor the powder to the caliber of the gun. The action of ordinary cannon powder was too sudden. The whole charge was consumed before the projectile had fairly started on its way, and the strain on the gun was terrific. Rodman compressed powder into disks that fitted the bore of the gun. The disks were an inch or two thick, and pierced with holes. With this arrangement, a minimum of powder surface was exposed at the beginning of combustion, but as the fire ate the holes larger (compare fig. 20f), the burning area actually in-

27

creased, producing a greater volume of gas as the projectile moved forward. Rodman thus laid the foundation for the "progressive burning" pellets of modern powders (fig. 20).

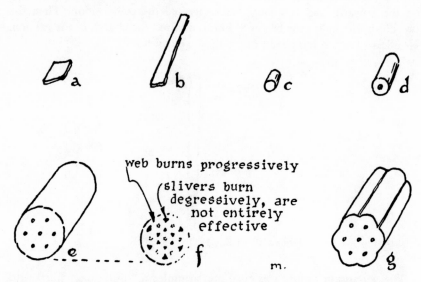

web burns progressively

slivers burn degressively, are not entirely effective

FIGURE 20—MODERN CANNON POWDER. A powder grain has the characteristics of an explosive only when it is confined. Modern *propellants* are low explosives (that is, relatively slow burning), but *projectiles* may be loaded with high explosive. a—Flake. b—Strip. c—Pellet. d—Single perforation. e—Standard 7-perforation. f—Burning grain of 7-perforation type. Ideally, the powder grain should burn progressively, with continuously increasing surface, the grain being completely consumed by the time the projectile leaves the bore. g—Walsh grain.

For a number of reasons General Rodman did not take his "perforated cake cartridge" beyond the experimental stage, and his "Mammoth" powder, such a familiar item in the powder magazines of the latter 1800's, was a compromise. As a block of wood burns steadier and longer than a quick-blazing pile of twigs, so the ¾-inch grains of mammoth powder gave a "softer" explosion, but one with more "push" and more uniform pressure along the bore of the gun.

It was in the second year of the Civil War that Alfred Nobel started the manufacture of nitroglycerin explosives in Europe. Smokeless powders came into use, the explosive properties of picric acid were discovered, and melanite, ballistite, and cordite appeared in the last quarter of the century, so that by 1890 nitrocellulose and nitroglycerin-base powders had generally replaced black powder as a propellant.

Still, black powder had many important uses. Its sensitivity to flame, high rate of combustion, and high temperature of explosion made it a very suitable igniter or "booster," to insure the complete ignition of the propellant. Further, it was the main element in such modern projectile fuzes as the ring fuze of the U. S. Field Artillery, which was long standard for bursts shorter than 25 seconds. This fuze was in the nose of the shell and consisted essentially of a plunger, primer, and rings grooved to hold a 9-inch train of compressed black powder. To set the fuze, the fuze man merely turned a movable ring to the proper time mark. Turning the zero mark toward the channel leading to the shell's bursting charge shortened

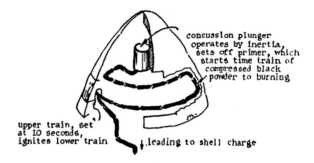

concussion plunger operates by inertia, sets off primer, which starts time train of compressed black powder to burning

upper train, set at 10 seconds, ignites lower train

leading to shell charge

FIGURE 21—MODERN POWDER TRAIN FUZE.

the burning distance of the train, while turning zero away from the channel, of course, did the opposite. When the projectile left the gun, the shock made the plunger ignite the primer (compare fig. 42e) and fire the powder train, which then burned for the set time before reaching the shell charge. It was a technical improvement over the tubular sheet-iron fuze of the Venetians, but the principle was about the same.

29

The Characteristics of Cannon

THE EARLY SMOOTHBORE CANNON

Soon after he found he could hurl a rock with his good right arm, man learned about trajectory—the curved path taken by a missile through the air. A baseball describes a "flat" trajectory every time the pitcher throws a hard, fast one. Youngsters tossing the ball to each other over a tall fence use "curved" or "high" trajectory. In artillery, where trajectory is equally important, there are three main types of cannon: (1) the flat trajectory gun, throwing shot at the target in relatively level flight; (2) the high trajectory mortar, whose shell will clear high obstacles and descend upon the target from above; and (3) the howitzer, an in-between piece of medium-high trajectory, combining the mobility of the fieldpiece with the large caliber of the mortar.

The Spaniard, Luis Collado, mathematician, historian, native of Lebrija in Andalusia, and, in 1592, royal engineer of His Catholic Majesty's Army in Lombardy and Piedmont, defined artillery broadly as "a machine of infinite importance." Ordnance he divided into three classes, admittedly following the rules of the "German masters, who were admired above any other nation for their founding and handling of artillery." Culverins and sakers (Fig. 23a) were guns of the first class, designed to strike the enemy from long range. The battering cannon (fig. 23b) were second class pieces; they were to destroy forts and walls and dismount the enemy's machines. Third class guns fired stone balls to break and sink ships and defend batteries from assault; such guns included the pedrero, mortar, and bombard (fig. 23c,d).

Collado's explanation of how the various guns were invented is perhaps naive, but nevertheless interesting: "Although the main intent of the inventors of this machine [artillery] was to fire and offend the enemy from both near and afar, since this offense must be in diverse ways it so happened that they formed various classes in this manner: they came to realize that men were not satisfied with the *espingardas* [small Moorish cannon], and for this reason the musket was made; and likewise the *esmeril* and the falconet. And although these fired longer shots, they

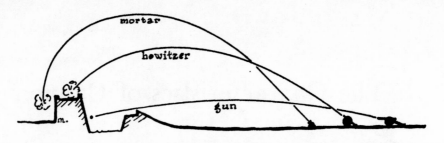

FIGURE 22—TRAJECTORIES.   Maximum range of eighteenth century guns was
                                        about 1 mile.

Guns could: Batter heavy construction with solid shot at long or short range;
                     destroy fort parapets and, by ricochet fire, dismount cannon; shoot
                     grape, canister, or bombs against massed personnel.

Mortars could: Reach targets behind obstructions; use high angle fire to shoot
                     bombs, destroying construction and personnel.

Howitzers could: Move more easily in the field than mortars; reach targets behind
                     obstructions by high angle fire; shoot larger projectiles than could
                     field guns of similar weight.

made the demisaker. To remedy a defect of that, the sakers were made,
and the demiculverins and culverins. While they were deemed sufficient
for making a long shot and striking the enemy from afar, they were of
little use as battering guns because they fire a small ball. So they determined to found a second kind of piece, wherewith, firing balls of much
greater weight, they might realize their intention. But discovering likewise that this second kind of piece was too powerful, heavy and costly for
batteries and for defense against assaults or ships and galleys, they made a
third class of piece, lighter in metal and taking less powder, to fire balls
of stone. These are the commonly called *cañones de pedreros.* All the
classes of pieces are different in range, manufacture and design. Even the
method of charging them is different."

It was most important for the artillerist to understand the different
classes of guns. As Collado quaintly phrased it, "he who ignores the pres-
ent lecture on this *arte* will, I assert, never do a good thing." Cannon
burst in the batteries every day because gunners were ignorant of how
the gun was made and what it was meant to do. Nor was such ignorance
confined to gunners alone. The will and whim of the prince who ordered
the ordnance or "the simple opinion of the unexpert founder himself,"
were the guiding principles in gun founding. "I am forced," wrote Collado, "to persuade the princes and advise the founders that the making of
artillery should always take into account the purpose each piece must
serve." This persuasion he undertook in considerable detail.

The first class of guns were the long-range pieces, comparatively "rich"
in metal. In the following table from Collado, the calibers and ranges for
most Spanish guns of this class are given, although as the second column
shows, at this period calibers were standardized only in a general way.

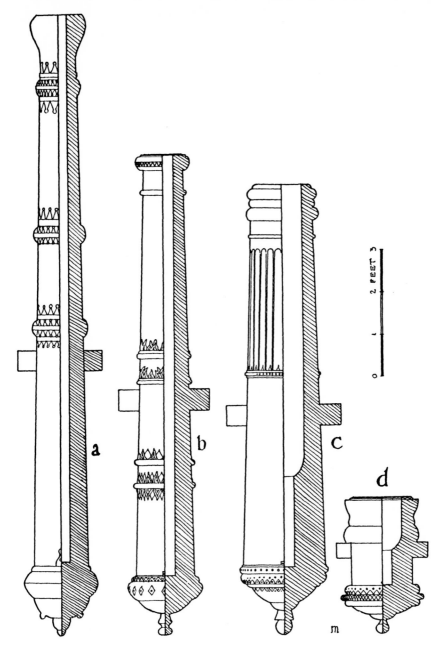

FIGURE 23—SIXTEENTH CENTURY SPANISH ARTILLERY. Taken from a 1592 manuscript, these drawings illustrate the three main classes of artillery used by Spain during the early colonial period in the New World. a—Culverin (Class 1). b—Cannon (Class 2). c—Pedrero (Class 3). d—Mortar (Class 3).

For translation where possible, and to list those which became the most popular calibers, we have added a final column. Most of the guns were probably of culverin length: 30- to 32-caliber.

Sixteenth century Spanish cannon of the first class

Name of gun	Weight of ball (pounds)	Length of gun (in calibers)	Range in yards		Popular caliber
			Point-blank	Maximum	
Esmeril....	½	....	208	750	.... ½-pounder esmeril.
Falconete..	1 to 2	....	......	......	.......1-pounder falconet.
Falcón....	3 to 4	....	417	2,500	.........3-pounder falcon.
Pasavolante	1 to 15	40 to 44	500	4,166	....6-pounder pasavolante.
Media sacre	5 to 7	....	417	3,750	.... 6-pounder demisaker.
Sacre.....	7 to 10	....	......	......	.........9-pounder saker.
Moyana...	8 to 10	shorter than saker	......	......	......9-pounder moyenne.
Media culebrina	10 to 18	....	833	5,000	..12-pounder demiculverin.
Tercio de culebrina	14 to 22	....	......	......	18-pounder third-culverin.
Culebrina..	20, 24, 25, 30, 40, 50	30 to 32	1,742	6,666	......24-pounder culverin
Culebrina real....	24 to 40	30 to 32	......	......	32-pounder culverin royal.
Doble culebrina	40 and up	30 to 32	......	......	48-pounder double culverin.

In view of the range Collado ascribes to the culverin, some remarks on gun performances are in order. "Greatest random" was what the old-time gunner called his maximum range, and random it was. Beyond point-blank range, the gunner was never sure of hitting his target. So with smoothbores, long range was never of great importance. Culverins, with their thick walls, long bores, and heavy powder charges, achieved distance; but second class guns like field "cannon," with less metal and smaller charges, ranged about 1,600 yards at a maximum, while the effective range was hardly more than 500. Heavier pieces, such as the French 33-pounder battering cannon, might have a point-blank range of 720 yards; at 200-yard range its ball would penetrate from 12 to 24 feet of earthwork, depending on how "poor and hungry" the earth was. At 130 yards a Dutch 48-pounder cannon put a ball 20 feet into a strong earth rampart, while from 100 yards a 24-pounder siege cannon would bury the ball 12 feet.

But generalizations on early cannon are difficult, for it is not easy to find two "mathematicians" of the old days whose ordnance lists agree. Spanish guns of the late 1500's do, however, appear to be larger in caliber than pieces of similar name in other countries, as is shown by comparing the culverins: the smallest Spanish *culebrina* was a 20-pounder, but the French great *coulevrine* of 1551 was a 15-pounder and the typical English culverin of that century was an 18-pounder. Furthermore, midway of the 1500's, Henry II greatly simplified French ordnance by holding his artillery down to the 33-pounder cannon, 15-pounder great culverin, 7½-pounder bastard culverin, 2-pounder small culverin, a 1-pounder falcon, and a ½-pounder falconet. Therefore, any list like the one following must have its faults:

Principal English guns of the sixteenth century

Name	Caliber (inches)	Length		Weight of gun (pounds)	Weight of shot (pounds)	Powder charge (pounds)
		Ft	In			
Rabinet................	1.0	....		300	0.3	0.18
Serpentine..............	1.5	....		400	.5	.3
Falconet................	2.0	3	9	500	1.0	.4
Falcon.................	2.5	6	0	680	2.0	1.2
Minion.................	3.5	6	6	1,050	5.2	3
Saker..................	3.65	6	11	1,400	6	4
Culverin bastard.........	4.56	8	6	3,000	11	5.7
Demiculverin...........	4.0	....		3,400	8	6
Basilisk................	5.0	....		4,000	14	9
Culverin...............	5.2	10	11	4,840	18	12
Pedrero................	6.0	....		3,800	26	14
Demicannon............	6.4	11	0	4,000	32	18
Bastard cannon..........	7.0	....		4,500	42	20
Cannon serpentine.......	7.0	....		5,500	42	25
Cannon................	8.0	....		6,000	60	27
Cannon royal...........	8.54	8	6	8,000	74	30

Like many gun names, the word "culverin" has a metaphorical meaning. It derives from the Latin *colubra* (snake). Similarly, the light gun called *áspide* or aspic, meaning "asp-like," was named after the venomous asp. But these digressions should not obscure the fact that both culverins and demiculverins were highly esteemed on account of their range and the effectiveness of fire. They were used for precision shooting such as building demolition, and an expert gunner could cut out a section of stone wall with these guns in short order.

As the fierce falcon hawk gave its name to the falcon and falconet, so the saker was named for the saker hawk; rabinet, meaning "rooster," was therefore a suitable name for the falcon's small-bore cousin. The 9-pounder saker served well in any military enterprise, and the *moyana* (or the French *moyenne*, "middle-sized"), being a shorter gun of saker caliber, was a good naval piece. The most powerful of the smaller pieces, however, was the *pasavolante*, distinguishable by its great length. It was between 40 and 44 calibers long! In addition, it had thicker walls than any other small caliber gun, and the combination of length and weight permitted an unusually heavy charge—as much powder as the ball weighed. A 6-pound lead ball was what the typical *pasavolante* fired; another gun of the same caliber firing an iron ball would be a 4-pounder. The point-blank range of this Spanish gun was a football field's length farther than either the falcon or demisaker.

In today's Spanish, *pasavolante* means "fast action," a phrase suggestive of the vicious impetuosity to be expected from such a small but powerful cannon. Sometimes it was termed a *drajón*, the English equivalent of which may be the drake, meaning "dragon"; but perhaps its most popular name in the early days was *cerbatana*, from Cerebus, the fierce three-headed dog of mythology. Strange things happen to words: a *cerbatana* in modern Spanish is a pea shooter.

Sixteenth century Spanish cannon of the second class

Spanish name	Weight of ball (pounds)	Translation
Quarto cañon	9 to 12	Quarter-cannon.
Tercio cañon	16	Third-cannon.
Medio cañon	24	Demicannon.
Cañon de abatir	32	Siege cannon.
Doble cañon	48	Double cannon.
Cañon de batería	60	Battering cannon.
Serpentino		Serpentine.
Quebrantamuro or lonbarda	70 to 90	Wallbreaker or lombard.
Basilisco	80 and up	Basilisk.

The second class of guns were the only ones properly called "cannon" in this early period. They were siege and battering pieces, and in some few respects were similar to the howitzers of later years. A typical Spanish cannon was only about two-thirds as long as a culverin, and the bore walls were thinner. Naturally, the powder charge was also reduced (half the ball's weight for a common cannon, while a culverin took double that amount).

The Germans made their light cannon 18 calibers long. Most Spanish siege and battering guns had this same proportion, for a shorter gun would not burn all the powder efficiently, "which," said Collado, "is a most grievous fault." However, small cannon of 18-caliber length were too short; the muzzle blast tended to destroy the embrasure of the parapet. For this reason, Spanish demicannon were as long as 24 calibers and the quarter-cannon ran up to 28. The 12-pounder quarter-cannon, incidentally, was "culverined" or reinforced so that it actually served in the field as a demiculverin.

The great weight of its projectile gave the double cannon its name. The warden of the Castillo at Milan had some 130-pounders made, but such huge pieces were of little use, except in permanent fortifications. It took a huge crew to move them, their carriages broke under the concentrated weight, and they consumed mountains of munitions. The lombard, which apparently originated in Lombardy, and the basilisk had the same disadvantages. The fabled basilisk was a serpent whose very look was fatal.

Its namesake in bronze was tremendously heavy, with walls up to 4 calibers thick and a bore up to 30 calibers long. It was seldom used by the Europeans, but the Turkish General Mustafa had a pair of basilisks at the siege of Malta, in 1565, that fired 150- and 200-pound balls. The 200-pounder gun broke loose as it was being transferred to a homeward bound galley and sank permanently to the bottom of the sea. Its mate was left on the island, where it became an object of great curiosity.

The third class of ordnance included the guns firing stone projectiles, such as the pedrero (or perrier, petrary, cannon petro, etc.), the mortars, and the old bombards like Edinburgh Castle's famous Mons Meg. Bars of wrought iron were welded together to form Meg's tube, and iron rings were clamped around the outside of the piece. In spite of many accidents, this coopering technique persisted through the fifteenth century. Mons Meg was made in two sections that screwed together, forming a piece 13 feet long and 5 tons in weight.

Pedreros (fig. 23c) were comparatively light. The foundryman used only half the metal he would put into a culverin, for the stone projectile weighed only a third as much as an iron ball of the same size, and the bore walls could therefore be comparatively thin. They were made in calibers up to 50-pounders. There was a chamber for the powder charge and little danger of the gun's bursting, unless a foolhardy fellow loaded it with an iron ball. The wall thicknesses of this gun are shown in Figure 24, where the inner circle represents the diameter of the chamber, the next arc the bore caliber, and the outer lines the respective diameters at chase, trunnions, and vent.

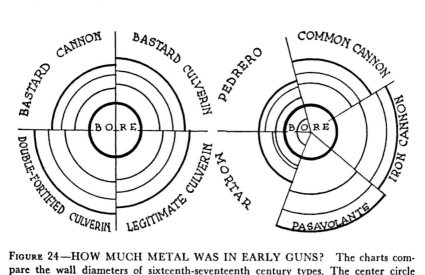

FIGURE 24—HOW MUCH METAL WAS IN EARLY GUNS? The charts compare the wall diameters of sixteenth-seventeenth century types. The center circle represents the bore, while the three outer arcs show the relative thickness of the bore wall at (1) the smallest diameter of the chase, (2) at the trunnions, and (3) at the vent. The small arc inside the bore indicates the powder chamber found in the pedrero and mortar.

Mortars (fig. 23d) were excellent for "putting great fear and terror in the souls of the besieged." Every night the mortars would play upon the town: "it keeps them in constant turmoil, due to the thought that some ball will fall upon their house." Mortars were designed like pedreros, except much shorter. The convenient way to charge them was with *saquillos* (small bags) of powder. "They require," said Collado, "a larger mouthful than any other pieces."

Just as children range from slight to stocky in the same family, there are light, medium, or heavy guns—all bearing the same family name. The difference lies in how the piece was "fortified"; that is, how thick the founder cast the bore walls. The English language has inelegantly descriptive terms for the three degrees of "fortification": (1) bastard, (2) legitimate, and (3) double-fortified. The thicker-walled guns used more powder. Spanish double-fortified culverins were charged with the full weight of the ball in powder; four-fifths that amount went into the legitimate, and only two-thirds for the bastard culverin. In a short culverin (say, 24 calibers long instead of 30), the gunner used 24/30 of a standard charge.

The yardstick for fortifying a gun was its caliber. In a legitimate culverin of 6-inch caliber, for instance, the bore wall at the vent might be one caliber (16/16 of the bore diameter) or 6 inches thick; at the trunnions it would be 10/16 or $4\frac{1}{8}$ inches, and at the smallest diameter of the chase, 7/16 or $2\frac{5}{8}$ inches. This table compares the three degrees of fortification used in Spanish culverins:

	Wall thickness in 8ths of caliber		
	Vent	Trunnion	Chase
Bastard culverin..........	7	5	3
Legitimate culverin.............................	8	$5\frac{1}{2}$	$3\frac{1}{2}$
Double-fortified culverin..................... .	9	$6\frac{1}{2}$	4

As with culverins, so with cannon. This is Collado's table showing the fortification for Spanish cannon:

	Wall thickness in 8ths of caliber		
	Vent	Trunnion	Chase
Cañon sencillo (light cannon)............... ..	6	$4\frac{1}{2}$	$2\frac{1}{2}$
Cañon común (common cannon)...................	7	5	$3\frac{1}{2}$
Canon reforzado (reinforced cannon)...............	8	$5\frac{1}{2}$	$3\frac{1}{2}$

Since cast iron was weaker than bronze, the walls of cast-iron pieces were even thicker than the culverins. Spanish iron guns were founded with 300 pounds of metal for each pound of the ball, and in lengths from 18 to 20 calibers. English, Irish, and Swedish iron guns of the period, Collado noted, had slightly more metal in them than even the Spaniards recommended.

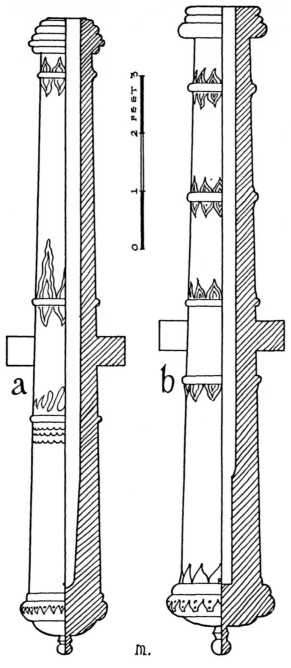

FIGURE 25—SIXTEENTH CENTURY CHAMBERED CANNON. a—"Bell-chambered" demicannon. b—Chambered demicannon.

Another way the designers tried to gain strength without loading the gun with metal was by using a powder chamber. A chambered cannon (fig. 25b) might be fortified like either the light or the common cannon, but it would have a cylindrical chamber about two-thirds of a caliber in diameter and four calibers long. It was not always easy, however, to get the powder into the chamber. Collado reported that many a good artillerist dumped the powder almost in the middle of the gun. When his ladle hit the mouth of the chamber, he thought he was at the bottom of the bore! The cylindrical chamber was somewhat improved by a cone-shaped taper, which the Spaniards called *encampanado* or "bell-chambered." A *cañon encampanado* (fig. 25a) was a good long-range gun, strong, yet light. But it was hard to cut a ladle for the long, tapered chamber.

Of all these guns, the reinforced cannon was one of the best. Since it had almost as much metal as a culverin, it lacked the defects of the chambered pieces. A 60-pounder reinforced cannon fired a convenient 55-pound ball, was easy to move, load, and clean, and held up well under any kind of service. It cooled quickly. Either cannon powder or fine powder (up to two-thirds the ball's weight) could be used in it. Reinforced cannon were an important factor in any enterprise, as King Philip's famed "Twelve Apostles" proved during the Flanders wars.

Fortification of sixteenth and seventeenth century guns

Spanish guns	Thickness of bore wall in 8ths of the caliber			English guns
	Vent	Trunnions	Chase	
Light cannon; bell-chambered cannon.......	6	4½	2½	........Bastard cannon.
Demicannon.............	6	5	3	
Common cannon; common siege cannon............	7	5	3½	
Light culverin; common battering cannon.........	7	5	3	.......Bastard culverin; legitimate cannon.
Common culverin; reinforced cannon.......	8	5½	3½	....Legitimate culverin; double-fortified cannon·
Legitimate culverin........	9	6½	4	Double-fortified culverin.
Cast-iron cannon..........	10	8	5	
Pasavolante..............	11½	8½	5½	

While there was little real progress in mobility until the days of Gustavus Adolphus, the wheeled artillery carriage seems to have been invented by the Venetians in the fifteenth century. The essential parts of the design were early established: two large, heavy cheeks or side pieces set on an axle and connected by transoms. The gun was cradled between the cheeks, the rear ends of which formed a "trail" for stabilizing and maneuvering the piece.

Wheels were perhaps the greatest problem. As early as the 1500's carpenters and wheelwrights were debating whether dished wheels were best. "They say," reported Collado, "that the [dished] wheel will never twist when the artillery is on the march. Others say that a wheel with spokes angled beyond the cask cannot carry the weight of the piece without twisting the spoke, so the wheel does not last long. I am of the same opinion, for it is certain that a perpendicular wheel will suffer more weight than the other. The defect of twisting under the pieces when on the march will be remedied by making the cart a little wider than usual." However, advocates of the dished wheel finally won.

SMOOTHBORES OF THE LATER PERIOD

From the guns of Queen Elizabeth's time came the 6-, 9-, 12-, 18-, 24-, 32-, and 42-pounder classifications adopted by Cromwell's government and used by the English well through the eighteenth century. On the Continent, during much of this period, the French were acknowledged leaders Louis XIV (1643-1715) brought several foreign guns into his ordnance, standardizing a set of calibers (4-, 8-, 12-, 16-, 24-, 32-, and 48-pounders) quite different from Henry II's in the previous century.

The cannon of the late 1600's was an ornate masterpiece of the foundryman's art, covered with escutcheons, floral relief, scrolls, and heavy moldings, the most characteristic of which was perhaps the banded muzzle (figs. 23b-c, 25, 26a-b), that bulbous bit of ornamentation which had been popular with designers since the days of the bombards. The flared or bell-shaped muzzle (figs. 23a, 26c, 27), did not supplant the banded muzzle until the eighteenth century, and, while the flaring bell is a usual characteristic of ordnance founded between 1730 and 1830, some banded-muzzle guns were made as late as 1746 (fig. 26a).

By 1750, however, design and construction were fairly well standardized in a gun of much cleaner line than the cannon of 1650. Although as yet there had been no sharp break with the older traditions, the shape and weight of the cannon in relation to the stresses of firing were becoming increasingly important to the men who did the designing.

Conditions in eighteenth century England were more or less typical: in the 1730's Surveyor-General Armstrong's formulae for gun design were hardly more than continuations of the earlier ways. His guns were about 20 calibers long, with these outside proportions:

1st reinforce $=2/7$ of the gun's length.
2d reinforce $=1/7$ plus 1 caliber.
chase $\quad =4/7$ less 1 caliber.

The trunnions, about a caliber in size, were located well forward (3/7 of the gun's length) "to prevent the piece from kicking up behind" when it was fired. Gunners blamed this bucking tendency on the practice of centering the trunnions on the *lower* line of the bore. "But what will not people do to support an old custom let it be ever so absurd?" asked John Müller, the master gunner of Woolwich. In 1756, Müller raised the trunnions to the *center* of the bore, an improvement that greatly lessened the strain on the gun carriage.

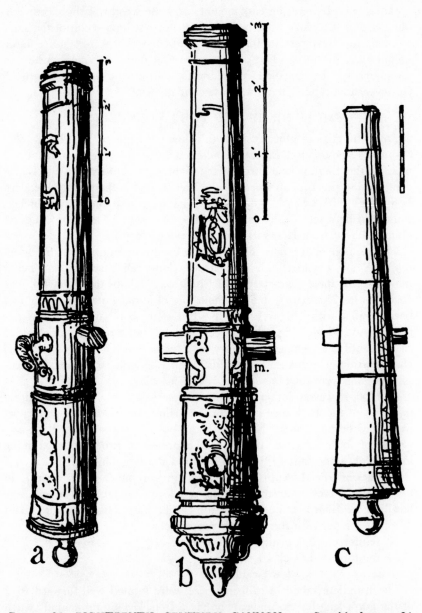

Figure 26—EIGHTEENTH CENTURY CANNON. a—Spanish bronze 24-pounder of 1746. b—French bronze 24-pounder of the early 1700's. c—English iron 6-pounder of the middle 1700's. The 6-pounder is part of the armament at Castillo de San Marcos.

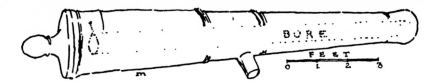

Figure 27—SPANISH 24-POUNDER CAST-IRON GUN (1693). Note the modern lines of this cannon, with its flat breech and slight muzzle swell.

The caliber of the gun continued to be the yardstick for "fortification" of the bore walls:

Vent	16	parts
End of 1st reinforce	14½	do
Beginning of second reinforce	13½	do
End of second reinforce	12½	do
Beginning of chase	11½	do
End of chase	8	do

For both bronze and iron guns, the above figures were the same, but for bronze, Armstrong divided the caliber into 16 parts; for iron it was only 14 parts. The walls of an iron gun thus were slightly thicker than those of a bronze one.

This eighteenth century cannon was a cast gun, but hoops and rings gave it the built-up look of the barrel-stave bombard, when hoops were really functional parts of the cannon. Reinforces made the gun look like "three frustums of cones joined together, so as the lesser base of the former is always greater than the greatest of the succeeding one." Ornamental fillets, astragals, and moldings, borrowed from architecture, increased the illusion of a sectional piece. Tests with 24-pounders of different lengths showed guns from 18 to 21 calibers long gave generally the best performance, but what was true for the 24-pounder was not necessarily true for other pieces. Why was the 32-pounder "brass battering piece" 6 inches longer than its 42-pounder brother? John Müller wondered about such inconsistencies and set out to devise a new system of ordnance for England.

Like many men before him, Müller sought to increase the caliber of cannon without increasing weight. He managed it in two ways: he modified exterior design to save on metal, and he lessened the powder charge to permit shortening and lightening the gun. Müller's guns had no heavy reinforces; the metal was distributed along the bore in a taper from powder chamber to muzzle swell. But realizing man's reluctance to accept new things, he carefully specified the location and size for each molding on his gun, protesting all the while the futility of such ornaments. Not until the last half of the next century were the experts well enough versed in metallurgy and interior ballistics to slough off all the useless metal.

So, using powder charges about one-third the weight of the projectile, Müller designed 14-caliber light field pieces and 15-caliber ship guns. His garrison and battering cannon, where weight was no great disadvantage,

43

were 18 calibers long. The figures in the table following represent the principal dimensions for the four types of cannon—all cast-iron except for the bronze siege guns. The first line in the table shows the length of the cannon. To proportion the rest of the piece, Müller divided the shot diameter into 24 parts and used it as a yardstick. The caliber of the gun, for instance, was 25 parts, or 25/24th of the shot diameter. The few other dimensions—thickness of the breech, length of the gun before the barrel began its taper, fortification at vent and chase—were expressed the same way.

	Field	Ship	Siege	Garrison
Length in calibers...........................	14	15	18	18
(Other proportions in 24ths of the shot diameter)				
Caliber..................................	25	25	25	25
Thickness of breech.......................	14	24	16	24
Length from breech to taper...............	39	49	40	49
Thickness at vent.........................	16	25	18	25
Thickness at muzzle.......................	8	12½	9	12½

The heaviest of Müller's garrison guns averaged some 172 pounds of iron for every pound of the shot, while a ship gun weighed only 146, less than half the iron that went into the sixteenth century cannon. And for a seafaring nation such as England, these were important things. Perhaps the opposite table will give a fair idea of the changes in English ordnance during the eighteenth century. It is based upon John Müller's lists of 1756; the "old" ordnance includes cannon still in use during Müller's time, while the "new" ordnance is Müller's own.

Windage in the English gun of 1750 was about 20 percent greater than in French pieces. The English ratio of shot to caliber was 20:21; across the channel it was 26:27. Thus, an English 9-pounder fired a 4.00-inch ball from a 4.20-inch bore; the French 9-pounder ball was 4.18 inches and the bore 4.34.

The English figured greater windage was both convenient and economical: windage, said they, ought to be just as thick as the metal in the gunner's ladle; standing shot stuck in the bore and unless it could be loosened with the ladle, had to be fired away and lost. John Müller brushed aside such arguments impatiently. With a proper wad over the shot, no dust or dirt could get in; and when the muzzle was lowered, said Müller, the shot "will roll out of course." Besides, compared with increased accuracy, the loss of a shot was trifling. Furthermore, with less room for the shot to bounce around the bore, the cannon would "not be spoiled so soon." Müller set the ratio of shot to caliber as 24:25.

In the 1700's cast-iron guns became the principal artillery afloat and ashore, yet cast bronze was superior in withstanding the stresses of firing. Because of its toughness, less metal was needed in a bronze gun than in a cast-iron one, so in spite of the fact that bronze is about 20 percent heavier

Calibers and lengths of principal eighteenth century English cannon

Caliber	Field Bronze/Iron		Ship Bronze		Ship Iron		Siege Bronze		Garrison Iron	
	Old	New	Old	New	Old	New	Old	New	Old	New
1½-pounder	.....	.	.....	.....	.....	.....	6'0"	.....	.....	.....
3-pounder	3'6"	3'3"	.....	3'6"	4'6"	3'6"	7'0"	.....	4'6"	4'2"
4-pounder	4'6"	4'1"	8'0"	4'4"	6'0"	4'4"	.....	.....	.....	.....
6-pounder	.....	4'8"	.....	5'0"	7'0"	5'0"	8'0"	.....	6'6"	5'3"
9-pounder	5'0"	5'1"	9'0"	5'6"	7'0"	5'6"	9'0"	.....	7'0"	6'0"
12-pounder	.....	5'10"	.....	6'4"	9'0"	6'4"	9'0"	6'7"	8'0"	6'7"
18-pounder	5'6"	6'5"	9'6"	7'0"	9'0"	7'0"	9'6"	7'6"	9'0"	7'6"
24-pounder	.....	.....	.....	7'6"	9'0"	7'6"	9'6"	8'4"	9'0"	8'4"
32-pounder	.....	.....	.....	7'10"	9'6"	.....	10'0"	9'2"	9'6"	9'2"
36-pounder	.....	.....	.....	.....	.....	.....	.....	9'6"	.....	.....
42-pounder	.....	.....	9'6"	8'4"	10'0"	8'4"	9'6"	10'0"	.....	10'0"
48-pounder	.....	.....	.....	8'6"	.....	8'6"	.....	10'6"	.....	.....

than iron, the bronze piece was usually the lighter of the two. For "position" guns in permanent fortifications where weight was no disadvantage, iron reigned supreme until the advent of steel guns. But non-rusting bronze was always preferable aboard ship or in seacoast forts.

Müller strongly advocated bronze for ship guns. "Notwithstanding all the precautions that can be taken to make iron Guns of a sufficient strength," he said, "yet accidents will sometimes happen, either by the mismanagement of the sailors, or by frosty weather, which renders iron very brittle." A bronze 24-pounder cost £156, compared with £75 for the iron piece, but the initial saving was offset when the gun wore out. The iron gun was then good for nothing except scrap at a farthing per pound, while the bronze cannon could be recast "as often as you please."

In 1740, Maritz of Switzerland made an outstanding contribution to the technique of ordnance manufacture. Instead of hollow casting (that is, forming the bore by casting the gun around a core), Maritz cast the gun solid, then drilled the bore, thus improving its uniformity. But although the bore might be drilled quite smooth, the outside of a cast-iron gun was always rough. Bronze cannon, however, could be put in the lathes to true up even the exterior. While after 1750 the foundries seldom turned out bronze pieces as ornate as the Renaissance culverins, a few decorations remained and many guns were still personalized with names in raised letters on the gun. Castillo de San Marcos has a 4-pounder "San Marcos," and, indeed, saints' names were not uncommon on Spanish ordnance. Other typical names were *El Espanto* (The Terror), *El Destrozo* (The Destroyer), *Generoso* (Generous), *El Toro* (The Bull), and *El Belicoso* (The Quarrelsome One).

In some instances, decoration was useful. The French, for instance, at one time used different shapes of cascabels to denote certain calibers; and even a fancy cascabel shaped like a lion's head was always a handy place for anchoring breeching tackle or maneuvering lines. The dolphins or handles atop bronze guns were never merely ornaments. Usually they were at the balance point of the gun; tackle run through them and hooked to the big tripod or "gin" lifted the cannon from its carriage.

GARRISON AND SHIP GUNS

Cannon for permanent fortifications were of various sizes and calibers, depending upon the terrain that had to be defended. At Castillo de San Marcos, for instance, the strongest armament was on the water front; lighter guns were on the land sector, an area naturally protected by the difficult terrain existing in the colonial period.

Before the Castillo was completed, guns were mounted only in the bastions or projecting corners of the fort. A 1683 inventory clearly shows that heaviest guns were in the San Agustín, or southeastern bastion, commanding not only the harbor and its entrance but the town of St. Augustine as well. San Pablo, the northwestern bastion, overlooked the land ap-

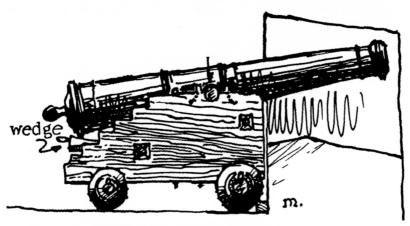

wedge

m.

FIGURE 28—EIGHTEENTH CENTURY SPANISH GARRISON GUN.

proach to the Castillo and the town gate; and, though its armament was lighter, it was almost as numerous as that in San Agustín. Bastion San Pedro to the southwest was within the town limits, and its few light guns were a reserve for San Pablo. The watchtower bastion of San Carlos overlooked the northern marshland and the harbor; its armament was likewise small. The following list details the variety and location of the ordnance:

Cannon mounted at Castillo de San Marcos in 1683

Location	No	Caliber	Class	Metal	Remarks
In the bastion of San Agustín	1	40-pounder	Cannon.....	Bronze	Carriage battered.
	1	18-pounder	..do.......	..do..	....New carriage.
	2	16-pounder	..do.......	Iron..	....Old carriages, wheels bad.
	1	12-pounder	..do.......	Bronze	....New carriage.
	1	12-pounder	..do.......	Iron..	.... do.
	1	8-pounder	..do.......	Bronze	.... Old carriage.
	1	7-pounder	..do.......	Iron..	....Carriage bad.
	1	4-pounder	..do.......	..do..	....New carriage.
	1	3-pounder	. do.......	Bronze	.... do.
In the bastion of San Pablo	1	16-pounder	Demicannon.	Iron..	.... Old carriage.
	1	10-pounder	Demiculverin.	Bronze	.... do.
	2	9-pounder	Cannon.....	Iron..	.... do.
	1	7-pounder	Demiculverin	Bronze	.... do.
	1	7-pounder	Cannon.....	Iron..	... Carriage bad.
	1	5-pounder	..do.......	..do..	... New carriage.
In the bastion of San Pedro	1	9-pounder	Cannon.....	Iron.	.... Old carriage.
	2	7-pounder	..do.......	..do..	....Carriage bad.
	2	5-pounder	..do.......	..do..	.... do.
	1	4-pounder	..do.......	Bronze	....Old carriage.
In the bastion of San Carlos	1	10-pounder	Cannon.....	Iron..	....Old carriage.
	1	5-pounder	..do.......	..do..	... New carriage.
	1	5-pounder	..do.......	Bronze	.. Good carriage.
	1	2-pounder	..do.......	Iron..	... New carriage.

The total number of Castillo guns in service at this date was 27, but there were close to a dozen unmounted pieces on hand, including a pair of pedreros. The armament was gradually increased to 70-odd guns as construction work on the fort made additional space available, and as other factors warranted more ordnance. Below is a summary of Castillo armament through the years:

Armament of Castillo de San Marcos, 1683-1834

Kind of gun	1683 Iron	1683 Bronze	1706 Iron	1706 Bronze	1740 Iron	1740 Bronze	1763 Iron	1763 Bronze	1765 Iron	1765 Bronze	1812 Iron	1812 Bronze	1834 Iron	1834 Bronze
2-pounder	1	..	..		..	..	..	..	..	..	..	..	..	..
3-pounder	..	1	..		2	3	..	..	..	..	..	..	..	..
4-pounder	1	1			5	1	..	..	..	..	1	..	..	..
5-pounder	4	1	26 guns from 4- to 10-pounders	8 guns from 2- to 16-pounders	15	1	..	..	.	..	..	..	..	..
6-pounder	..	..			5	..	..	..	..	1	..	..	3	..
7-pounder	4	1			5	2	..	..	..	..	..	..	..	..
8-pounder	..	1			11	1	5	11	..	..	1	..	..	..
3½-in. carronade	..	..			..	..	..	..	..	4	..	..	..	..
9-pounder	3	..			..	..	..	..	..	..	..	..	..	..
10-pounder	1	1			..	..	6	..	..	..	..	..	..	..
12-pounder	1	1	..		..	..	13	..	7	..	2	..	..	..
15-pounder	..	..	..		6	..	..	..	..	..	..	..	..	..
16-pounder	3	..	..		..	..	2	1	..	..	8	..	..	..
18-pounder	..	1	..	..	4	1	7	..	..	..	..	..	4	..
24-pounder	..	..	..	..	2	..	7	..	32	..	10	..	5	..
33-pounder	..	..	..	..	..	1	..	..	..	..	..	..	..	..
36-pounder	..	..	..	1	..	..	..	1	..	..	..	.	..	..
40-pounder	..	1	..	..	..	..	..	..	..	..	..	..	..	..
24-pounder field howitzer	..	..	..	..	..	..	..	..	..	..	..	..	2	2
6-in. howitzer	..	..	..	..	..	..	..	..	..	..	..	2	..	2
8-in. howitzer	..	..	..	..	..	..	..	..	..	2	..	..	..	..
Small mortar	..	..	..	..	..	..	..	18	..	20	..	..	..	..
6-in. mortar	..	..	..	..	..	..	..	..	..	..	..	1	..	1
9-in. mortar	..	..	..	..	..	..	..	..	..	..	..	1	..	..
10-in. mortar	..	..	..	..	..	..	..	..	..	..	..	..	..	1
Large mortar	..	..	..	..	..	..	..	6	..	1	..	..	..	..
Stone mortar	2	..	..	..	..	..	..	..	..	..	..	3	..	..
Total	20	9	26	9	55	10	40	37	39	24	26	8	14	6
Grand total	29		35		65		77		63		34		20	

This tabulation reflects contemporary conditions quite clearly. The most serious invasions of Spanish Florida took place during the first half of the eighteenth century, precisely the time when the Castillo armament was strongest. While most of the guns were in battery condition, the table does have some pieces rated only fair and may also include a few unserviceables. Colonial isolation meant that ordnance often served longer than the normal 1,200-round life of an iron piece. A usual failure was the development of cracks around the vent or in the bore. Sometimes a muzzle blew off. The worst casualties of the 1702 siege came from the bursting of an iron 16-pounder which killed four and seriously wounded six men. At that period, incidentally, culverins were the only guns with the range to reach the harbor bar some 3,000 yards away.

Although when the Spanish left Florida to Britain in 1763 they took serviceable cannon with them, two guns at Castillo de San Marcos National Monument today appear to be seventeenth century Spanish pieces. Most of the 24- and 32-pounder garrison cannon, however, are English-founded, after the Armstrong specifications of the 1730's, and were part of the British armament during the 1760's. Amidst the general confusion and shipping troubles that attended the British evacuation in 1784, some ordnance seems to have been left behind, to remain part of the defenses until the cession to the United States in 1821.

The Castillo also has some interesting United States guns, including a pair of early 24-pounder iron field howitzers (c. 1777-1812). During the 1840's the United States modernized Castillo defenses by constructing a water battery in the moat behind the sea wall. Many of the guns for that battery are extant, including 8-inch Columbiads, 32-pounder cannon, 8-inch seacoast and garrison howitzers. St. Augustine's Plaza even boasts a converted 32-pounder rifle.

FIGURE 29—VAUBAN'S MARINE CARRIAGE (c. 1700).

Garrison and ship carriages were far different from field, siege, and howitzer mounts, while mortar beds were in a separate class entirely. Basic proportions for the carriage were obtained by measuring (1) the distance from trunnion to base ring of the gun, (2) the diameter of the base ring, and (3) the diameter of the second reinforce ring. The result was a quadrilateral figure that served as a key in laying out the carriage to fit the gun. Cheeks, or side pieces, of the carriage were a caliber in thickness, so the bigger the gun, the more massive the mount.

A 24-pounder cheek would be made of timber about 6 inches thick. The Spaniards often used mahogany. At Jamestown, in the early 1600's, Capt. John Smith reported the mounting of seven "great pieces of ordnance upon new carriages of cedar," and the French colonials also used this material. British specifications in the mid-eighteenth century called for cheeks and transoms of dry elm, which was very pliable and not likely to split; but some carriages were made of young oak, and oak was stand-

ard for United States garrison carriages until it was replaced by wrought-
iron after the Civil War.

For a four-wheeled English carriage of 1750, height of the cheek was
4¾ diameters of the shot, unless some change in height had to be made
to fit a gun port or embrasure. To prevent cannon from pushing shutters
open when the ship rolled in a storm, lower tier carriages let the muzzle
of the gun, when fully elevated, butt against the sill over the gun port.

On the eighteenth century Spanish garrison carriage (fig. 28), no bolts
were threaded; all were held either by a key run through a slot in the
foot of the bolt, or by bradding the foot over a decorative washer. Compared with American mounts of the same type (figs. 30 and 31), the
Spanish carriage was considerably more complicated, due partly to the
greater amount of decorative ironwork and partly to the design of the
wooden parts which, with their carefully worked mortises, required a
craftsman's skill. The cheek of the Spanish carriage was a single great
plank. English and American construction called for a built-up cheek of
several planks, cleverly jogged or mortised together to prevent starting
under the strain of firing.

FIGURE 30—ENGLISH GARRISON CARRIAGE (1756). By substituting wood-
en wheels for the cast-iron ones, this carriage became a standard naval gun carriage.

Müller furnished specifications for building truck (four-wheeled) carriages for 3- to 42-pounders. Aboard ship, of course, the truck carriage
was standard for almost everything except the little swivel guns and the
mortars.

Carriage trucks (wheels), unless they were made of cast iron, had iron
thimbles or bushings driven into the hole of the hub, and to save the wood
of the axletree, the spindle on which the wheel revolved was partly protected by metal. The British put copper on the *bottom* of the spindle;
Spanish and French designers put copper on the *top*, then set iron "axletree bars" into the bottom. These bars strengthened the axletree and re-
sisted wear at the spindle.

A 24-pounder fore truck was 18 inches in diameter. Rear trucks were
16 inches. The difference in size compensated for the slope in the gun
platform or deck—a slope which helped to check recoil. Aboard ship,
where recoil space was limited, the "kick" of the gun was checked by a

heavy rope called a breeching, shackled to the side of the vessel (see fig. 11). Ship carriages of the two- or four-wheel type (fig. 31), were used through the War between the States, and there was no great change until the advent of automatic recoil mechanisms made a stationary mount possible.

wedge

FIGURE 31—U. S. NAVAL TRUCK CARRIAGE (1866).

With garrison carriages, however, changes came much earlier. In 1743, Fort William on the Georgia coast had a pair of 18-pounders mounted upon "curious moving Platforms" which were probably similar to the traversing platforms standardized by Gribeauval in the latter part of the century. United States forts of the early 1800's used casemate and barbette carriages (fig. 10) of the Gribeauval type, and the traversing platform of these mounts made training (aiming the gun right or left) comparatively easy.

Training the old truck carriage had been heavy work for the handspikemen, who also helped to elevate or depress the gun. Maximum elevation or depression was about 15° each way—about the same as naval guns used during the Civil War. If one quoin was not enough to secure proper depression, a block or a second quoin was placed below the first. But before the gunner depressed a smoothbore below zero elevation, he had to put either a wad or a grommet over the ball to keep it from rolling out.

Ship and garrison cannon were not moved around on their carriages. If the gun had to be taken any distance, it was dismounted and chained under a sling wagon or on a "block carriage," the big wheels of which easily rolled over difficult terrain. It was not hard to dismount a gun: the keys locking the cap squares were removed, and then the gin was rigged and the gun hoisted clear of the carriage.

A typical garrison or ship cannon could fire any kind of projectile, but solid shot, hot shot, bombs, grape, and canister were in widest use. These guns were flat trajectory weapons, with a point-blank range of about 300 yards. They were effective—that is, fairly accurate—up to about half a mile, although the maximum range of guns like the Columbiad of the nineteenth century, when elevation was not restricted by gun port confines, approached the 4-mile range claimed by the Spanish for the six-

51

teenth century culverin. The following ranges of United States ordnance in the 1800's are not far different from comparable guns of earlier date.

Ranges of United States smoothbore garrison guns of 1861

Caliber	Elevation	Range in yards
18-pounder siege and garrison.............................	5° 0″	1,592
24-pounder siege and garrison.............................	5° 0″	1,901
32-pounder seacoast......................................	5° 0″	1,922
42-pounder seacoast......................................	5° 0″	1,955
8-inch Columbiad..	27°30″	4,812
10-inch Columbiad.......................................	39°15″	5,654
12-inch Columbiad.......................................	39° 0″	5,506

Ranges of United States naval smoothbores of 1866

Caliber	Point-blank range in yards	Elevation	Range in yards
32-pounder of 42 cwt........................	313	5°	1,756
8-inch of 63 cwt............................	330	5°	1,770
IX-inch shell gun...........................	350	15°	3,450
X-inch shell gun...........................	340	11°	3,000
XI-inch shell gun..........................	295	15°	2,650
XV-inch shell gun.........................	300	7°	2,100

Ranges of United States naval rifles in 1866

Caliber	Elevation	Range in yards
20-pounder Parrott...................................	15°	4,400
30-pounder Parrott...................................	25°	6,700
100-pounder Parrott.............	25°	7,180

In accuracy and range the rifle of the 1860's far surpassed the smoothbores, but such tremendous advances were made in the next few decades with the introduction of new propellants and steel guns that the performances of the old rifles no longer seem remarkable. In the eighteenth century, a 24-pounder smoothbore could develop a muzzle velocity of about 1,700 feet per second. The 12-inch rifled cannon of the late 1800's had a muzzle velocity of 2,300 foot-seconds. In 1900, the Secretary of the Navy proudly reported that the new 12-inch guns for *Maine*-class battleships produced a muzzle velocity of 2,854 foot-seconds, using an 850-pound projectile and a charge of 360 pounds of smokeless powder. Such statistics elicit a chuckle from today's artilleryman.

SIEGE CANNON

Field counterpart of the garrison cannon was the siege gun—the "battering cannon" of the old days, mounted upon a two-wheeled siege or

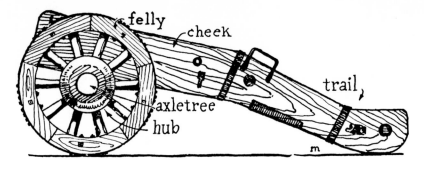

FIGURE 32—SPANISH EIGHTEENTH CENTURY SIEGE CARRIAGE.

"traveling" carriage that could be moved about in field terrain. Whereas the purpose of the garrison cannon was to destroy the attacker and his matériel, the siege cannon was intended to destroy the fort. Calibers ranged from 3- to 42-pounders in eighteenth century English tables, but the 18- and 24-pounders seem to have been the most widely used for siege operations.

The siege carriage closely resembled the field gun carriage, but was much more massive, as may be seen from these comparative figures drawn from eighteenth century English specifications:

24-pounder field carriage		24-pounder siege carriage
9 feet long..................	Length of cheek..............	........13 feet.
4.5 inches..................	Thickness of cheek...........	....5.8 inches.
50 inches..................	Wheel diameter..............	....58 inches.
6x8x68 inches..............	Axletree...................	7x9x81 inches.

Heavy siege guns were elevated with quoins, and elevation was restricted to 12° or less, which was about the same as United States siege carriages permitted in 1861. It was considered ample for these flat trajectory pieces.

Both field and siege carriages were pulled over long distances by lifting the trail to a horse- or ox-drawn limber; a hole in the trail transom seated on an iron bolt or pintle on the two-wheeled limber. Some late eighteenth century field and siege carriages had a second pair of trunnion holes a couple of feet back from the regular holes, and the cannon was shifted to the rear holes where the weight was better distributed for traveling. The United States siege carriage of the 1860's had no extra trunnion holes, but a "traveling bed" was provided where the gun was cradled in position 2 or 3 feet back of its firing position. A well-drilled gun crew could make the shift very rapidly, using a lifting jack, a few rollers, blocks, and chocks. When there was danger of straining or breaking the gun carriage, however, massive block carriages, sling carts, or wagons were used to carry the guns.

Sling wagons were of necessity used for transport in siege operations when the guns were to be mounted on barbette (traversing platform) carriages (fig. 10). Emplacing the barbette carriage called for construction of a massive, level subplatform, but it also eliminated the old need for the gunner to chalk the location of his wheels in order to return his gun to the proper firing position after each shot.

The Federal sieges of Forts Pulaski and Sumter were highly complicated engineering operations that involved landing tremendously heavy ordnance (the 300-pounder Parrott weighed 13 tons) through the surf, moving the big guns over very difficult terrain and, in some cases, building roads over the marshes and driving foundation piles for the gun emplacements.

The heavy caliber Parrotts trained on Fort Sumter were in batteries from 1,750 to 4,290 yards distant from their target. They were very accurate, but their endurance was an uncertain factor. The notorious "Swamp Angel," for instance, burst after 36 rounds.

FIELD CANNON

The field guns were the mobile pieces that could travel with the army and be brought quickly into firing position. They were lighter in weight

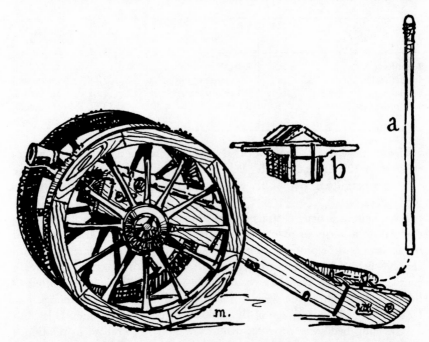

Figure 33—SPANISH 4-POUNDER FIELD CARRIAGE (c. 1788). This carriage, designed on the "new method," employed a handscrew instead of a wedge for elevating the piece. a—The handspike was inserted through eye-bolts in the trail. b—The ammunition locker held the cartridges.

54

than any other type of flat trajectory weapon. To achieve this lightness the designers had not only shortened the guns, but thinned down the bore walls. In the eighteenth century, calibers ran from the 3- to the 24-pounder, mounted on comparatively light, two-wheeled carriages. In addition, there was the 1½-pounder (and sometimes the light 3- or 6-pounder) on a "galloper" carriage—a vehicle with its trail shaped into shafts for the horse. The elevating-screw mechanism was early developed for field guns, although the heavier pieces like the 18- and 24-pounders were still elevated by quoins as late as the early 1800's.

In the Castillo collection are parts of early United States field carriages little different from Spanish carriages that held a score of 4-pounders in the long, continuous earthwork parapet surrounding St. Augustine in the eighteenth century. The Spanish mounts were a little more complicated in construction than English or American carriages, but not much. Spanish pyramid-headed nails for securing ironwork were not far different from the diamond- and rose-headed nails of the English artificer.

Each piece of hardware on the carriage had its purpose. Gunner's tools were laid in hooks on the cheeks. There were bolts and rings for the lines when the gun had to be moved by manpower in the field. On the trail transom, pintle plates rimmed the hole that went over the pintle on the limber. Iron reinforced the carriage at weak points or where the wood was subject to wear. Iron axletrees were common by the late 1700's.

For training the field gun, the crew used a special handspike quite different from the garrison handspike. It was a long, round staff, with an iron handle bolted to its head (fig. 33a). The trail transom of the carriage held two eyebolts, into which the foot of the spike was inserted. A lug fitted into an offset in the larger eyebolt so that the spike could not twist. With the handspike socketed in the eyebolts, lifting the trail and laying the gun was easy.

The single-trail carriage (fig. 13) used so much during the middle 1800's was a remarkable simplification of carriage design. It was also essential for guns like the Parrott rifles, since the thick reinforce on the breech of an otherwise slender barrel would not fit the older twin-trail carriage. The single, solid "stock" or trail eliminated transoms, for to the sides of the stock itself were bolted short, high cheeks, humped like a camel to cradle the gun so high that great latitude in elevation was possible. The elevating screw was threaded through a nut in the stock, right under the big reinforce of the gun.

While the larger bore siege Parrotts were not noted for long serviceability, Parrott field rifles had very high endurance. As for performance, see the following table:

Caliber	Weight of gun (pounds)	Type of projectile	Projectile weight (pounds)	Elevation	Range (yards)	Smoothbore of same caliber
10-pounder....	890	Shell.......	9.75	5°	2,000	3-pounder.
		..do.......	9.75	20°	5,000	
20-pounder....	1,750	..do.......	18.75	5°	2,100	6-pounder.
		..do.......	18.75	15°	4,400	
30-pounder....	4,200	..do.......	29.00	15°	4,800	9-pounder.
		..do.......	29.00	25°	6,700	
		Long shell...	101.00	15°	4,790	
		..do.......	101.00	25°	6,820	
		Hollow shot..	80.00	25°	7,180	
		.do.......	80.00	35°	8,453	

Amazingly enough, these ranges were obtained with about the same amount of powder used for the smoothbores of similar caliber: the 10-pounder Parrott used only a pound of powder; the 20-pounder used a two-pound charge; and the 30-pounder, 3¼ pounds!

HOWITZERS

The howitzer was invented by the Dutch in the seventeenth century to throw larger projectiles (usually bombs) than could the field pieces, in a high trajectory similar to the mortar, but from a lighter and more mobile weapon. The wide-purpose efficiency of the howitzer was appreciated almost at once, and it was soon adopted by all European armies. The weapon owed its mobility to a rugged, two-wheeled carriage like a field carriage, but with a relatively short trail that permitted the wide arc of elevation needed for this weapon.

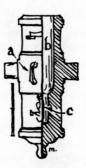

FIGURE 34—SPANISH 6-INCH HOWITZER (1759-88). This bronze piece was founded during the reign of Charles III and bears his shield. a—Dolphin, or handle. b—Bore. c—Powder chamber.

English howitzers of the 1750's were of three calibers: 5.8-, 8-, and 10-inch, but the 10-incher was so heavy (some 50 inches long and over 3,500

pounds) that it was quickly discarded. Müller deplored the superfluous
weight of these pieces and developed 6-, 8-, 10, and 13-inch howitzers in

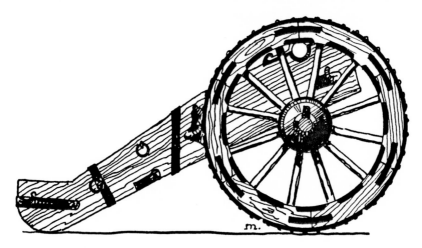

FIGURE 35—ENGLISH 8-INCH "HOWITZ" CARRIAGE (1756). The short
trail enabled greater latitude in elevating the howitzer.

which, by a more calculated distribution of the metal, he achieved much
lighter weapons. Müller's howitzers survived in the early 6- to 10-inch
pieces of United States artillery and one fine little 24-pounder of the late
eighteenth century happens to be among the armament of Castillo de San
Marcos, along with some early nineteenth century howitzers. The British,
incidentally, were the first to bring this type gun to Florida. None appeared on the Castillo inventory until the 1760's.

In addition to the very light and therefore easily portable mountain
howitzer used for Indian warfare, United States artillery of 1850 included
12-, 24-, and 32-pounder field, 24-pounder and 8-inch siege and garrison,
and the 10-inch seacoast howitzer. The Navy had a 12-pounder heavy
and a 24-pounder, to which were added the 12- and 24-pounder Dahlgren
rifled howitzers of the Civil War period. Such guns were often used in
landing operations. The following table gives some typical ranges:

Ranges of U. S. Howitzers in the 1860's

Caliber	Elevation	Range i. yards
10-inch seacoast..	5°	1,650
8-inch siege..	12°30″	2,280
24-pounder naval...	5°	1,270
12-pounder heavy naval.....................................	5°	1,085
20-pounder Dahlgren rifled.................................	5°	1,960
12-pounder Dahlgren rifled.................................	5°	1,770

57

From earliest times the usefulness of the mortar as an arm of the artillery has been clearly recognized. Up until the 1800's the weapon was usually made of bronze, and many mortars had a fixed elevation of 45°,

FIGURE 36—ENGLISH MORTAR ON ELEVATING BED (1740).

which in the sixteenth century was thought to be the proper elevation for maximum range of any cannon. In the 1750's Müller complained of the stupidity of English artillerists in continuing to use fixed-elevation mortars, and the Spanish made a *mortero de plancha,* or "plate" mortar (fig. 37), as late as 1788. Range for such a fixed-elevation weapon was varied by using more or less powder, as the case required. But the most useful mortar, of course, had trunnions and adjustable elevation by means of quoins.

The mortar was mounted on a "bed"—a pair of wooden cheeks held together by transoms. Since a bed had no wheels, the piece was trans-

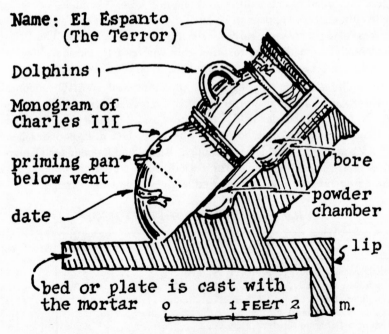

FIGURE 37—SPANISH 5-INCH BRONZE MORTAR (1788).

ported on a mortar wagon or sling cart. In the battery, the mortar was generally bedded upon a level wooden platform; aboard ship, it was a revolving platform, so that the piece could be quickly aimed right or left. The mortar's weight, plus the high angle of elevation, kept it pretty well in place when it was fired, although English artillerists took the additional precaution of lashing it down.

The mortar did not use a wad, because a wad prevented the fuze of the shell from igniting. To the layman, it may seem strange that the shell was never loaded with the fuze toward the powder charge of the gun. But the fuze was always toward the muzzle and away from the blast, a practice which dated from the early days when mortars were discharged by "double firing": the gunner lit the fuze of the shell with one hand and the priming of the mortar with the other. Not until the late 1600's did the method of letting the powder blast ignite the fuze become general. It was a change that greatly simplified the use of the arm and, no doubt, caused the mortarman to heave a sigh of relief.

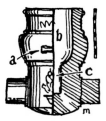

FIGURE 38—SPANISH 10-INCH BRONZE MORTAR (1759-88). a—Dolphin, or handle. b—Bore. c—Powder chamber.

Most mortars were equipped with dolphins, either singly or in pairs, which were used for lifting the weapon onto its bed. Often there was a little bracketed cup—a priming pan—under the vent, a handy gadget that saved spilling a lot of powder at the almost vertical breech. As with other bronze cannon, mortars were embellished with shields, scrolls, names, and other decoration.

About 1750, the French mortar had a bore length 1½ diameters of the shell; in England, the bore was 2 diameters for the smaller calibers and 3 for the 10- and 13-inchers. The extra length added a great deal of weight to the English mortars: the 13-inch weighed 25 hundredweight, while the French equivalent weighed only about half that much. Müller complained that mortar designers slavishly copied what they saw in other guns. For instance, he said, the reinforce was unnecessary; it ". . . overloads the Mortar with a heap of useless metal, and that in a place where the least strength is required, yet as if this unnecessary metal was not sufficient, they add a great projection at the mouth, which serves to no other purpose than to make the Mortar top-heavy. The mouldings are likewise jumbled together, without any taste or method, tho' they are taken from architecture." Field mortars in use during Müller's time included 4.6-, 5.8-, 8-, 10-, and 13-inch "land" mortars and 10- and 13-inch "sea" mortars. Müller, of course, redesigned them.

Figure 39—COEHORN MORTAR. The British General Oglethorpe used 20 coehorns in his 1740 bombardment of St. Augustine. These small mortars were also used extensively during the War Between the States.

The small mortars called coehorns (fig. 39) were invented by the famed Dutch military engineer, Baron van Menno Coehoorn, and used by him in 1673 to the great discomfit of French garrisons. Oglethorpe had many of them in his 1740 bombardment of St. Augustine when the Spanish, trying to translate coehorn into their own tongue, called them *cuernos de vaca*—"cow horns." They continued in use through the U.S. Civil War, and some of them may still be seen in the battlefield parks today.

Bombs and carcasses were usual for mortar firing, but stone projectiles remained in use as late as 1800 for the pedrero class (fig. 43). Mortar projectiles were quite formidable; even in the sixteenth century missiles weighing 100 or more pounds were not uncommon, and the 13-inch mortar of 1860 fired a 200-pound shell. The larger projectiles had to be whipped up to the muzzle with block and tackle.

Figure 40—THE "DICTATOR." This huge 13-inch mortar was used by the Federal artillery in the bombardment of Petersburg, Va., 1864-65.

In the last century, the bronze mortars metamorphosed into the great cast-iron mortars, such as "The Dictator," that mammoth Federal piece used against Petersburg, Va. Wrought-iron beds with a pair of rollers were built for them. In spite of their high trajectory, mortars could range well over a mile, as witness these figures for United States mortars of the 1860's, firing at 45° elevation:

Ranges of U. S. Mortars in 1861

Caliber	Projectile weight (pounds)	Range (yards)
8-inch siege.......	45	1,837
10-inch siege....	90	2,100
12-inch seacoast..	200	4,625
13-inch seacoast...	200	4,325

At the siege of Fort Pulaski in 1862, however, General Gillmore complained that the mortars were highly inaccurate at mile-long range. On this point, John Müller would have nodded his head emphatically. A hundred years before Gillmore's complaint, Müller had argued that a range of something less than 1,500 yards was ample for mortars or, for that matter, all guns. "When the ranges are greater," said Müller, "they are so uncertain, and it is so difficult to judge how far the shell falls short, or exceeds the distance of the object, that it serves to no other purpose than to throw away the Powder and shell, without being able to do any execution."

PETARDS

"Hoist with his own petard," an ancient phrase signifying that one's carefully laid scheme has exploded, had truly graphic meaning in the old days when everybody knew what a petard was. Since the petard fired no projectile, it was hardly a gun. Roughly speaking, it was nothing but an iron bucket full of gunpowder. The petardier would hang it on a gate, something like hanging your hat on a nail, and blast the gate open by firing the charge.

Small petards weighed about 50 pounds; the large ones, around 70 pounds. They had to be heavy enough to be effective, yet light enough for a couple of men to lift up handily and hang on the target. The bucket part was packed full of the powder mixture, then a 2½-inch-thick board was bolted to the rim in order to keep the powder in and the air out. An iron tube fuze was screwed into a small hole in the back or side of the weapon. When all was ready, the petardiers seized the two handles of the petard and carried it to the troublesome door. Here they set a screw, hung the explosive instrument upon it, lit the fuze, and "retired."

Petards were used frequently in King William's War of the 1680's to force the gates of small German towns. But on a well-barred, double gate the small petard was useless, and the great petard would break only the fore part of such a gate. Furthermore, as one would guess, hanging a petard was a hazardous occupation; it went out of style in the early 1700's.

Projectiles

There are four different types of artillery projectiles which, in one form or another, have been used since very early times:

(1) Battering projectiles (solid shot).
(2) Exploding shells.
(3) Scatter shot (case or canister, grape, shrapnel).
(4) Incendiary and chemical projectiles.

SOLID SHOT

At Havana, Cuba, in the early days, there was an abundance of round stones lying around, put there by Mother Nature. Artillerists at Havana never lacked projectiles. Stone balls, cheap to manufacture, relatively light and therefore well suited to the feeble construction of early ordnance, were in general use for large caliber cannon in the fourteenth century. There were experiments along other lines such as those at Tournay in the 1330's with long, pointed projectiles. Lead-coated stones were fairly popular, and solid lead balls were used in some small pieces, but the stone ball was more or less standard.

Cast-iron shot had been introduced by 1400, and, with the improvement of cannon during that century, iron shot gradually replaced stone. By the end of the 1500's stone survived for use only in the pedreros, murtherers, and other relics of the earlier period. Iron shot for the smoothbore was a solid, round shot, cast in fairly accurate molds; the mold marks that invariably show on all cannonballs were of small importance, for the ball did not fit the bore tightly. After casting, shot were checked with a ring gauge (fig. 41)—a hoop through which each ball had to pass. The Spanish term for this tool is very descriptive: *pasabala*, "ball-passer."

Shot was used mainly in the flat-trajectory cannon. The small caliber guns fired nothing but shot, for small sizes of the other type projectiles were not effective. Shot was the prescription when the situation called for "great accuracy, at very long range," and penetration. Fired at ships, a shot was capable of breaching the planks (at 100-yard range a 24-pounder shot would penetrate 4½ feet of "sound and hard" oak). With a fair aim

63

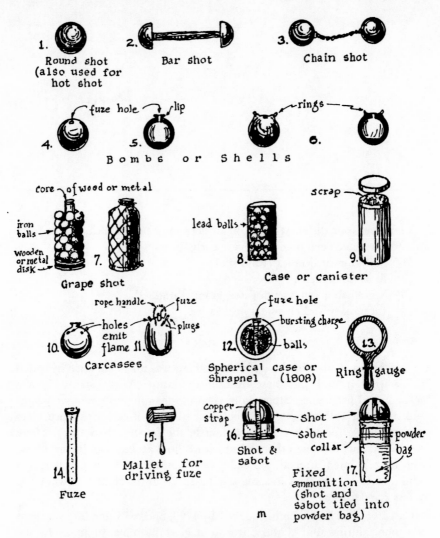

1. Round shot (also used for hot shot

2. Bar shot

3. Chain shot

4. fuze hole

5. lip

Bombs or Shells

6. rings

core of wood or metal

iron balls

Wooden or metal disk

7. Grape shot

scrap

lead balls

8.

9. Case or canister

rope handle — fuze

holes emit flame

plugs

10. 11. Carcasses

fuze hole

bursting charge

balls

12. Spherical case or Shrapnel (1808)

13. Ring gauge

14. Fuze

15. Mallet for driving fuze

copper strap

Shot

sabot

16. Shot & sabot

collar

Shot

powder bag

Fixed 17. ammunition (shot and sabot tied into powder bag)

m

FIGURE 41—EIGHTEENTH CENTURY PROJECTILES. (Not to scale.)

at the waterline, a gunner could sink or seriously damage a vessel with a few rounds. On ironclad targets like the *Monitor* and *Merrimac,* however, round shot did little more than bounce; it took the long, armor-piercing rifle projectile to force the development of the tremendously thick plate of modern times.

Round shot was very useful for knocking out enemy batteries. The gunner put his cannon on the flank of the hostile guns and used ricochet firing so that the ball, just clearing the defense wall, would bounce among the enemy guns, wound the crews, and break the gun carriages. In the destruction of fort walls, shot was essential. After dismounting the enemy pieces, the siege guns moved close enough to batter down the walls. The procedure was not as haphazard as it sounds. Cannon were brought as close as possible to the target, and the gunner literally cut out a low section with gunfire so that the wall above tumbled down into the moat and made a ramp right up to the breach. Firing at the upper part of the wall defeated its own purpose, for the rubble brought down only protected the foundation area, and the breach was so high that assault troops had to use ladders.

The most effective bombardment of Castillo de San Marcos occurred during the 1740 siege, and shot did the most damage. The heaviest English siege cannon were 18-pounders, over 1,000 yards from the fort. Spanish Engineer Pedro Ruiz de Olano reported that the balls did not penetrate the massive main walls more than a foot and a half, but the parapets, being only 3 feet thick, suffered considerable damage. Some of the old parapets, Engineer Ruiz said, "have been demolished, and the new ones have suffered very much owing to their recent construction." (He meant that the new mortar had not sufficiently hardened.) Ruiz was not deceived about what would happen if hostile batteries were able to get closer; in such case, he thought, the enemy "will no doubt succeed in destroying the parapets and dismounting the guns."

Variations of round shot were bar shot and chain shot (fig. 41), two or more projectiles linked together for simultaneous firing. Bar shot appears in a Castillo inventory of 1706, and like chain shot, was for specialized work like cutting a ship's rigging. There is one apocryphal tale, however, about an experiment with chain shot as anti-personnel missiles: instead of charging a single cannon with the two balls, two guns were used, side by side. The ball in one gun was chained to the ball in the other. The projectiles were to fly forth, stretching the long chain between them, mowing down a sizeable segment of the enemy. Instead, the chain wrapped the gun crews in a murderous embrace; one gun had fired late.

EXPLOSIVE SHELLS

The word "bomb" comes to us from the French, who derived it from the Latin. But the Romans got it originally from the Greek *bombos*, meaning a deep, hollow sound. "Bombard" is a derivation. Today bomb is pronounced "balm," but in the early days it was commonly pronounced "bum." The modern equivalent of the "bum" is an HE shell.

The first recorded use of explosive shells was by the Venetians in 1376. Their bombs were hemispheres of stone or bronze, joined together with hoops and exploded by means of a primitive powder fuze. Shells filled with explosive or incendiary mixtures were standard for mortars, after 1550, but they did not come into general use for flat-trajectory weapons until early in the nineteenth century, whereafter the term "shell" gradually won out over "bomb."

In any event, this projectile was one of the most effective ever used in the smoothbore against earthworks, buildings, and for general bombardment. A delayed action shell, diabolically timed to roll amongst the ranks with its fuze burning, was calculated to "disorder the stoutest men," since they could not know at what awful instant the bomb would burst.

A bombshell was simply a hollow, cast-iron sphere. It had a single hole where the powder was funneled in—full, but not enough to pack too tightly when the fuze was driven in. Until the 1800's, the larger bombs were not always smooth spheres, but had either a projecting neck, or collar, for the fuze hole or a pair of rings at each side of the hole for easier handling (fig. 41). In later years, however, such projections were replaced by two "ears," little recesses beside the fuze hole. A pair of tongs (something like ice tongs) seized the shell by the ears and lifted it up to the gun bore.

During most of the eighteenth century, shells were cast thicker at the base than at the fuze hole on the theory that they were (1) better able to resist the shock of firing from the cannon and (2) more likely to fall with the heavy part underneath, leaving the fuze uppermost and less liable to extinguishment. Müller scoffed at the idea of "choaking" a fuze, which, he said, burnt as well in water as in any other element. Furthermore, he preferred to use shells "everywhere equally thick, because they would then burst into a greater number of pieces." In later years, the shells were scored on the interior to ensure their breaking into many fragments.

FUZES

The eighteenth century fuze was a wooden tube several inches long, with a powder composition tamped into its hole much like the nineteenth century fuze (fig. 42c). The hole was only a quarter of an inch in diam-

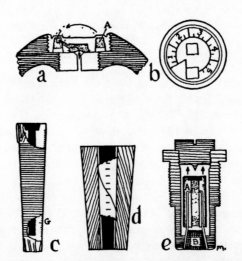

FIGURE 42—NINETEENTH CENTURY PROJECTILE FUZES. a—Cross-section of Bormann fuze. b—Top of Bormann fuze. c—Wooden fuze for spherical shell. d—Wood-and-paper fuze for spherical shell. e—Percussion fuze.

eter, but the head of the fuze was hollowed out like a cup, and "mealed" (fine) powder, moistened with "spirits of wine" (alcohol), was pressed into the hollow to make a larger igniting surface. To time the fuze, a cannoneer cut the cylinder at the proper length with his fuze-saw, or drilled a small hole (G) where the fire could flash out at the right time. Some English fuzes at this period were also made by drawing two strands of a quickmatch into the hole, instead of filling it with powder composition. The ends of the match were crossed into a sort of rosette at the head of the fuze. Paper caps to protect the powder composition covered the heads of these fuzes and had to be removed before the shell was put into the gun.

Bombs were not filled with powder very long before use, and fuzes were not put into the projectiles until the time of firing. To force the fuze into the hole of the shell, the cannoneer covered the fuze head with tow, put a fuze-setter on it, and hammered the setter with a mallet, "drifting" the fuze until the head stuck out of the shell only $\frac{2}{10}$ of an inch. If the fuze had to be withdrawn, there was a fuze extractor for the job. This tool gripped the fuze head tightly, and turning a screw slowly pulled out the fuze.

Wooden tube fuzes were used almost as long as the spherical shell. A United States 12-inch mortar fuze (fig. 42c), 7 inches long and burning 49 seconds, was much like the earlier fuze. During the 1800's, however, other types came into wide use.

The conical paper-case fuze (fig. 42d), inserted in a metal or wooden plug that fitted the fuze hole, contained composition whose rate of burning was shown by the color of the paper. A black fuze burned an inch every 2 seconds. Red burned 3 seconds, green 4, and yellow 5 seconds per inch. Paper fuzes were 2 inches long, and could be cut shorter if necessary. Since firing a shell from a 24-pounder to burst at 2,000 yards meant a time flight of 6 seconds, a red fuze would serve without cutting, or a green fuze could be cut to $1\frac{1}{2}$ inches. Sea-coast fuzes of similar type were used in the 15-inch Rodmans until these big smoothbores were finally discarded sometime after 1900.

The Bormann fuze (fig. 42a), the quickest of the oldtimers to set, was used for many years by the U.S. Field Artillery in spherical shell and shrapnel. Its pewter case, which screwed into the shell, contained a time ring of powder composition (A). Over this ring the top of the fuze case was marked in seconds. To set the fuze, the gunner merely had to cut the case at the proper mark—at four for 4 seconds, three for 3 seconds, and so on—to expose the ring of powder to the powder blast of the gun. The ring burned until it reached the zero end and set off the fine powder in the center of the case; the powder flash then blew out a tin plate in the bottom of the fuze and ignited the shell charge. Its short burning time (about 6 seconds) made the Bormann fuze obsolete as field gun ranges increased. The main trouble with this fuze, however, was that it did not always ignite!

The percussion fuze was an extremely important development of the nineteenth century, particularly for the long-range rifles. The shock of impact caused this fuze to explode the shell at almost the instant of striking. Percussion fuzes were made in two general types: the front fuze, for the nose of an elongated projectile; and the base fuze, at the center of the projectile base. The base fuze was used with armor-piercing projectiles where it was desirable to have the shell penetrate the target for some distance before bursting. Both types were built on the same principles.

A Hotchkiss front percussion fuze (fig. 42e) had a brass case which screwed into the shell. Inside the case was a plunger (A) containing a priming charge of powder, topped with a cap of fulminate. A brass wire at the base of the plunger was a safety device to keep the cap away from a sharp point at the top of the fuze until the shell struck the target. When the gun was fired, the shock of discharge dropped a lead plug (B) from the base of the fuze into the projectile cavity, permitting the plunger to drop to the bottom of the fuze and rest there, held by the spread wire, while the shell was in flight. Upon impact, the plunger was thrown forward, the cap struck the point and ignited the priming charge, which in turn fired the bursting charge of the shell.

SCATTER PROJECTILES

When one of our progenitors wrathfully seized a handful of pebbles and flung them at the flock of birds in his garden, he discovered the principle of the scatter projectile. Perhaps its simplest application was in the stone mortar (fig. 43). For this weapon, round stones about the size of a

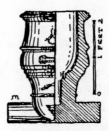

FIGURE 43—SPANISH 16-INCH PEDRERO (1788). This mortar fired baskets of stones.

man's fist (and, by 1750, hand grenades) were dumped into a two-handled basket and let down into the bore. This primitive charge was used at close range against personnel in a fortification, where the effect of the descending projectiles would be uncommonly like a short but severe barrage of over-sized hailstones. There were 6,000 stones in the ammunition inventory for Castillo de San Marcos in 1707.

One of the earliest kinds of scatter projectiles was case shot, or canister, used at Constantinople in 1453. The name comes from its case, or can, usually metal, which was filled with scrap, musket balls, or slugs (fig. 41). Somewhat similar, but with larger iron balls and no metal case, was grape shot, so-called from the grape-like appearance of the clustered balls. A stand of grape in the 1700's consisted of a wooden disk at the base of a short wooden rod that served as the core around which the balls stood (fig. 41). The whole assembly was bagged in cloth and reinforced with a net of heavy cord. In later years grape was made by bagging two or three tiers of balls, each tier separated by an iron disk. Grape could disable men at almost 900 yards and was much used during the 1700's. Eventually, it was almost replaced by case shot, which was more effective at shorter ranges (400 to 700 yards). Incidentally, there were 2,000 sacks of grape at the Castillo in 1740, more than any other type projectile.

Spherical case shot (fig. 41) was an attempt to carry the effectiveness of grape and canister beyond its previous range, by means of a bursting shell. It was the forerunner of the shrapnel used so much in World War I and was invented by Lt. Henry Shrapnel, of the British Army, in 1784. There had been previous attempts to produce a projectile of this kind, such as the German Zimmerman's "hail shot" of 1573—case shot with a bursting charge and a primitive time fuze—but Shrapnel's invention was the first air-bursting case shot which, in technical words, "imparted directional velocity" to the bullets it contained. Shrapnel's new shell was first used against the French in 1808, but was not called by its inventor's name until 1852.

INCENDIARIES AND CHEMICAL PROJECTILES

Incendiary missiles, such as buckets or barrels filled with a fiercely burning composition, had been used from earliest times, long before cannon. These crude incendiaries survived through the 1700's as, for instance, the flaming cargoes of fire ships that were sent amidst the enemy fleet. But in the year 1672 there appeared an iron shell called a carcass (fig. 41), filled with pitch and other materials that burned at intense heat for about 8 minutes. The flame escaped through vents, three to five in number, around the fuze hole of the shell. The carcass was standard ammunition until smoothbores went out of use. The United States ordnance manual of 1861 lists carcasses for 12-, 18-, 24-, 32-, and 42-pounder guns as well as 8-, 10-, and 13-inch mortars.

During the late 1500's, the heating of iron cannon balls to serve as incendiaries was suggested, but not for another 200 years was the idea successfully carried out. Hot shot was nothing but round shot, heated to a red glow over a grate or in a furnace. It was fired from cannon at such inflammable targets as wooden ships or powder magazines. During the siege of Gibraltar in 1782, the English fired and destroyed a part of Spain's fleet with hot shot; and in United States seacoast forts shot furnaces were standard equipment during the first half of the 1800's. The little shot furnace at Castillo de San Marcos National Monument was built during the 1840's;

a giant furnace of 1862 still remains at Fort Jefferson National Monument. Few other examples are left.

Loading hot shot was not particularly dangerous. After the powder charge was in the gun with a dry wad in front of it, another wad of wet straw, or clay, was put into the barrel. When the cherry-red shot was rammed home, the wet wad prevented a premature explosion of the charge. According to the *Ordnance Manual*, the shot could cool in the gun without setting off the charge! Hot shot was superseded, about 1850, by Martin's shell, filled with molten iron.

The smoke shell appeared in 1681, but was never extensively used. Similarly, a form of gas projectile, called a "stink shell," was invented by a Confederate officer during the Civil War. Because of its "inhumanity," and probably because it was not thought valuable enough to offset its propaganda value to the enemy, it was not popular. These were the beginnings of the modern chemical shells.

In connection with chemical warfare, it is of interest to review the Hussite siege of Castle Karlstein, near Prague, in the first quarter of the fifteenth century. The Hussites emplaced 46 small cannon, 5 large cannon, and 5 catapults. The big guns would shoot once or twice a day, and the little ones from six to a dozen rounds.

Marble pillars from Prague churches furnished the cannonballs. Many projectiles for the catapults, however, were rotting carcasses and other filth, hurled over the castle walls to cause disease and break the morale of the besieged. But the intrepid defenders neutralized these "chemical bursts" with lime and arsenic. After firing 10,930 cannonballs, 932 stone fragments, 13 fire barrels, and 1,822 tons of filth, the Hussites gave up.

FIXED AMMUNITION

In early days, due partly to the roughly made balls, wads were very important as a means of confining the powder and increasing its efficiency. Wads could be made of almost any suitable material at hand, but perhaps straw or hay ones were most common. The hay was first twisted into a 1-inch rope, then a length of the rope was folded together several times and finally rolled up into a short cylinder, a little larger than the bore. After the handier sabots came into use, however, wads were needed only to keep the ball from rolling out when the muzzle was down, or for hot shot firing.

Gunners early began to consolidate ammunition for easier and quicker loading. For instance, after the powder charge was placed in a bag, the next logical step was to attach the wad and the cannonball to it, so that loading could be made in one simple operation—pushing the single round into the bore (fig. 48). Toward that end, the sabot or "shoe" (fig. 41) took the place of the wad. The sabot was a wooden disk about the same diameter as the shot. It was secured to the ball with a pair of metal straps to make "semi-fixed" ammunition; then, if the neck of the powder bag were tied around the sabot, the result was one cartridge, containing powder, sabot, and ball, called "fixed" ammunition. Fixed ammunition was usual for the lighter field pieces by the end of the 1700's, while the bigger guns used "semi-fixed."

In transportation, cartridges were protected by cylinders and caps of strong paper. Sabots were sometimes made of paper, too, or of compressed wood chips, to eliminate the danger of a heavy, unbroken sabot falling amongst friendly troops. A big mortar sabot was a lethal projectile in itself!

ROCKETS

Today's rocket projectiles are not exactly new inventions. About the time of artillery's beginning, the military fireworker came into the business of providing pyrotechnic engines of war; later, his job included the spectacular fireworks that were set off in celebration of victory or peace.

Artillery manuals of very early date include chapters on the manufacture and use of fireworks. But in making war rockets there was no marked progress until the late eighteenth century. About 1780, the British Army in India watched the Orientals use them; and within the next quarter century William Congreve, who set about the task of producing a rocket that would carry an incendiary or explosive charge as far as 2 miles, had achieved such promising results that English boats fired rocket salvos against Boulogne in 1806. The British Field Rocket Brigade used rockets effectively at Liepsic in 1812—the first time they appeared in European land warfare. They were used again 2 years later at Waterloo. The warheads of such rockets were cast iron, filled with black powder and fitted with percussion fuzes. They were fired from trough-like launching stands, which were adjustable for elevation.

Rockets seem to have had a demoralizing effect upon untrained troops, and perhaps their use by the English against raw American levies at Bladenburg, in 1814, contributed to the rout of the United States forces and the capture of Washington. They also helped to inspire Francis Scott Key. Whether or not he understands the technical characteristics of the rocket, every schoolboy remembers the "rocket's red glare" of the National Anthem, wherein Key recorded his eyewitness account of the bombardment of Fort McHenry. The U.S. Army in Mexico (1847) included a rocket battery, and, indeed, war rockets were an important part of artillery resources until the rapid progress of gunnery in the latter 1800's made them obsolescent.

Tools

Gunner's equipment was numerous. There were the tompion (a lid that fitted over the muzzle of the gun to keep wind and weather out of the bore) and the lead cover for the vent; water buckets for the sponges and passing boxes for the powder; scrapers and tools for "searching" the bore to find dangerous cracks or holes; chocks for the wheels; blocks and rollers, lifting jacks, and gins for moving guns; and drills and augers for clearing the vent (figs. 17, 44). But among the most important tools for everyday firing were the following:

The sponge was a wooden cylinder about a foot long, the same diameter as the shot, and covered with lambskin. Like all bore tools, it was mounted on a long staff; after being dampened with water, it was used for cleaning the bore of the piece after firing. Essentially, sponging made sure there were no sparks in the bore when the new charge was put in. Often the sponge was on the opposite end of the rammer, and sometimes, instead of being lambskin-covered, the sponge was a bristle brush.

The wormer was a double screw, something like a pair of intertwined corkscrews, fixed to a long handle. Inserted in the gun bore and twisted, it seized and drew out wads or the remains of cartridge bags stuck in the gun after firing. Worm screws were sometimes mounted in the head of the sponge, so that the piece could be sponged and wormed at the same time.

The ladle was the most important of all the gunner's tools in the early years, since it was not only the measure for the powder but the only way to dump the powder in the bore at the proper place. It was generally made of copper, the same gauge as the windage of the gun; that is, the copper was just thick enough to fit between ball and bore.

Essentially, the ladle is merely a scoop, a metal cylinder secured to a wooden disk on a long staff. But before the introduction of the powder cartridge, cutting a ladle to the right size was one of the most important accomplishments a gunner had to learn. Collado, that Spanish mathematician of the sixteenth century, used the culverin ladle as the master pattern (fig. 45). It was 4½ calibers long and would carry exactly the weight of the ball in powder. Ladles for lesser guns could be proportioned (that is, shortened) from the master pattern.

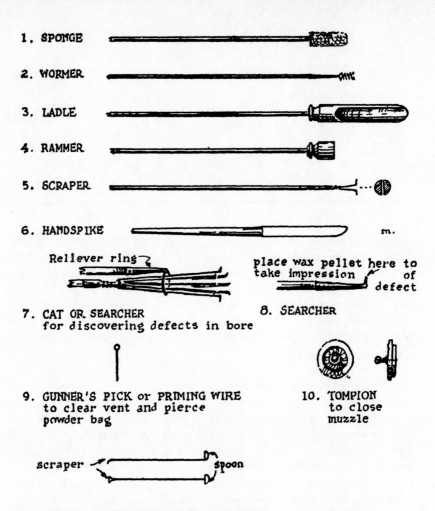

1. SPONGE

2. WORMER

3. LADLE

4. RAMMER

5. SCRAPER

6. HANDSPIKE   m.

Reliever ring

place wax pellet here to
take impression   of
                  defect

7. CAT OR SEARCHER   8. SEARCHER
   for discovering defects in bore

9. GUNNER'S PICK or PRIMING WIRE   10. TOMPION
   to clear vent and pierce   to close
   powder bag   muzzle

scraper   spoon

11. BORE SCRAPER-AND-SPOON for mortar

Figure 44—EIGHTEENTH CENTURY GUNNER'S EQUIPMENT. (Not to scale.)

The ladle full of powder was pushed home in the bore. Turning the handle dumped the charge, which then had to be packed with the rammer. As powder charges were lessened in later years, the ladle was shortened; by 1750, it was only three shot diameters long. With cartridges, the ladle was no longer needed for loading the gun, but it was still handy for withdrawing the round.

The rammer was a wooden cylinder about the same diameter and length as the shot. It pushed home the powder charge, the wad, and the shot. As a precaution against faulty or double loading, marks on the rammer handle showed the loaders when the different parts of the charge were properly seated.

74

The gunner's pick or priming wire was a sharp pointed tool resembling a common ice pick blade. It was used to clear the vent of the gun and to pierce the powder bag so that flame from the primer could ignite the charge.

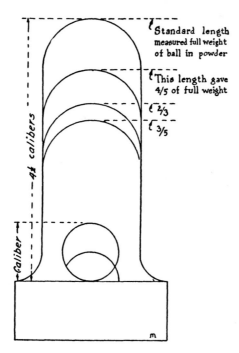

Standard length
measured full weight
of ball in powder

This length gave
4/5 of full weight

2/3

3/5

4½ calibers

Caliber

FIGURE 45—SIXTEENTH CENTURY PATTERN FOR GUNNER'S LADLE.

Handspikes were big pinch bars to manhandle cannon. They were used to move the carriage and to lift the breech of the gun so that the elevating quoin or screw might be adjusted. They were of different types (figs. 33a, 44), but were essentially 6-foot-long wooden poles, shod with iron. Some of them, like the Marsilly handspike (fig. 11), had rollers at the toe so that the wheelless rear of the carriage could be lifted with the handspike and rolled with comparative ease.

The gunner's quadrant (fig. 46), invented by Tartaglia about 1545, was an aiming device so basic that its principle is still in use today. The instrument looked like a carpenter's square, with a quarter-circle connecting the two arms. From the angle of the square dangled a plumb bob. The gunner laid the long arm of the quadrant in the bore of the gun, and the line of the bob against the graduated quarter-circle showed the gun's angle of elevation.

The addition of the quadrant to the art of artillery opened a whole new field for the mathematicians, who set about compiling long, complicated, and jealously guarded tables for the gunner's guidance. But the theory was simple: since a cannon at 45° elevation would fire *ten* times farther than it would when the barrel was level (at zero° elevation), the quadrant should be marked into *ten* equal parts; the range of the gun would therefore increase by *one-tenth* each time the gun was elevated to the next mark on the quadrant. In other words, the gunner could get the range he wanted simply by raising his piece to the proper mark on the instrument.

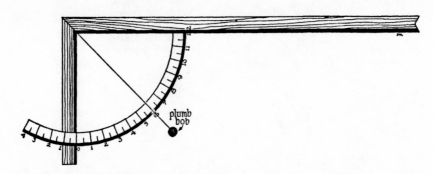

FIGURE 46—SEVENTEENTH CENTURY GUNNER'S QUADRANT. The long end of the quadrant was laid in the bore of the cannon. The plumb bob indicated the degree of elevation on the scale.

Collado explained how it worked in the 1590's. "We experimented with a culverin that fired a 20-pound iron ball. At point-blank the first shot ranged 200 paces. At 45-degree elevation it shot ten times farther, or 2,000 paces. . . . If the point-blank range is 200 paces, then elevating to the *first* position, or a tenth part of the quadrant, will gain 180 paces more, and advancing another point will gain so much again. It is the same with the other points up to the elevation of 45 degrees; each one gains the same 180 paces." Collado admitted that results were not always consistent with theory, but it was many years before the physicists understood the effect of air resistance on the trajectory of the projectile.

Sights on cannon were usually conspicuous by their absence in the early days. A dispart sight (an instrument similar to the modern infantry rifle sight), which compensated for the difference in diameter between the breech and the muzzle, was used in 1610, but the average artilleryman still aimed by sighting over the barrel. The Spanish gunner, however, performed an operation that put the bore parallel to the gunner's line of sight, and called it "killing the *vivo*" (*matar el vivo*). How *vivo* affected aiming is easily seen: with its bore level, a 4-pounder falconet ranged 250 paces. But when the *top of the gun* was level, the bore was slightly elevated and the range almost doubled to 440 paces.

To "kill the *vivo*," you first had to find it. The gunner stuck his pick into the vent down to the bottom of the bore and marked the pick to show the depth. Next he took the pick to the muzzle, stood it up in the bore, and marked the height of the muzzle. The difference between the two marks, with an adjustment for the base ring (which was higher than the vent), was the *vivo*. A little wedge of the proper size, placed under the breech, would then eliminate the troublesome *vivo*.

During the first half of the 1700's Spanish cannon of the "new invention" were made with a notch at the top of the base ring and a sighting button on the muzzle, and these features were also adopted by the French. But they soon went out of use. There was some argument, as late as the 1750's, about the desirability of casting the muzzle the same size as the base ring, so that the sighting line over the gun would always be parallel to the bore; but, since the gun usually had to be aimed higher than the objective, gunners claimed that a fat muzzle hid their target!

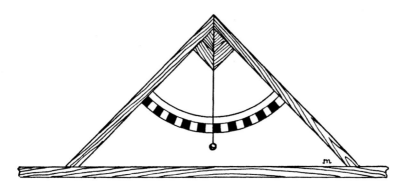

FIGURE 47—SEVENTEENTH CENTURY GUNNER'S LEVEL. This tool was useful in many ways, but principally for finding the line of sight on the barrel of the gun.

Common practice for sighting, as late as the 1850's, was to find the center line at the top of the piece, mark it with chalk or filed notches, and use it as a sighting line. To find this center line, the gunner laid his level (fig. 47) first on the base ring, then on the muzzle. When the instrument was level atop these rings, the plumb bob was theoretically over the center line of the cannon. But guns were crudely made, and such a line on the outside of the piece was not likely to coincide exactly with the center line of the bore, so there was still ample opportunity for the gunner to exercise his "art." Nonetheless the marked lines did help, for the gunner learned by experiment how to compensate for errors.

Fixed rear sights came into use early in the 1800's, and tangent sights (graduated rear sights) were in use during the War Between the States. The trunnion sight, a graduated sight attached to the trunnion, could be used when the muzzle had to be elevated so high that it blocked the gunner's view of the target.

Naval gunnery officers would occasionally order all their guns trained at the same angle and elevated to the same degree. The gunner might not even see his target. While with the crude traversing mechanism of the early 1800's the gunners may not have laid their pieces too accurately, at least it was a step toward the indirect firing technique of later years which was to take full advantage of the longer ranges possible with modern cannon. Use of tangent and trunnion sights brought gunnery further into the realm of mathematical science; the telescopic sight came about the middle of the nineteenth century; gunners were developing into technicians whose job was merely to load the piece and set the instruments as instructed by officers in fire control posts some distance away from the gun.

The Practice of Gunnery

The oldtime gunner was not only an artist, vastly superior to the average soldier, but, when circumstances permitted, he performed his wizardry with all due ceremony. Diego Ufano, Governor of. Antwerp, watched a gun crew at work about 1500:

"The piece having arrived at the battery and being provided with all needful materials, the gunner and his assistants take their places, and the drummer is to beat a roll. The gunner cleans the piece carefully with a dry rammer, and in pulling out the said rammer gives a dab or two to the mouth of the piece to remove any dirt adhering." (At this point it was customary to make the sign of the cross and invoke the intercession of St. Barbara.)

"Then he has his assistant hold the sack, valise, or box of powder, and filling the charger level full, gives a slight movement with the other hand to remove any surplus, and then puts it into the gun as far as it will go. Which being done, he turns the charger so that the powder fills the breech and does not trail out on the ground, for when it takes fire there it is very annoying to the gunner." (And probably to the gentleman holding the sack.)

"After this he will take the rammer, and, putting it into the gun, gives two or three good punches to ram the powder well in to the chamber, while his assistant holds a finger in the vent so that the powder does not leap forth. This done, he takes a second charge of powder and deposits it like the first; then puts in a wad of straw or rags which will be well packed to gather up all the loose powder. This having been well seated with strong blows of the rammer, he sponges out the piece.

"Then the ball, well cleaned by his assistant, since there is danger to the gunner in balls to which sand or dirt adhere, is placed in the piece without forcing it till it touches gently on the wad, the gunner being careful not to hold himself in front of the gun, for it is silly to run danger without reason. Finally he will put in one more wad, and at another roll of drums the piece is ready to fire."

Maximum firing rate for field pieces in the early days was eight rounds an hour. It increased later to 100 rounds a day for light guns and 30 for

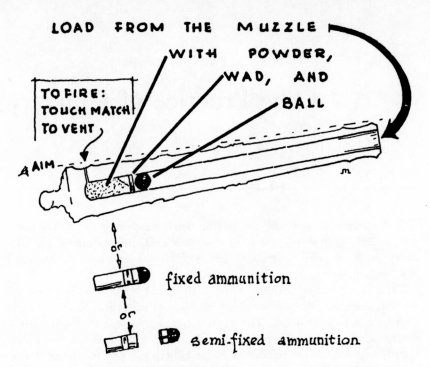

FIGURE 48—LOADING A CANNON. Muzzle-loading smoothbore cannon were used for almost 700 years.

heavy pieces. (Modern nonautomatic guns can fire 15 rounds per minute.) After about 40 rounds the gun became so hot it was unsafe to load, where-upon it was "refreshed" with an hour's rest.

Approved aiming procedure was to make the first shot surely short, in order to have a measurement of the error. The second shot would be at greater elevation, but also cautiously short. After the third round, the gun-ner could hope to get hits. Beginners were cautioned against the desire to hit the target at the first shot, for, said a celebrated artillerist, ". . . you will get overs and cannot estimate how much over."

As gunners gradually became professional soldiers, gun drills took on a more military aspect, as these seventeenth century commands show:

1. Put back your piece.
2. Order your piece to load.
3. Search your piece.
4. Sponge your piece.
5. Fill your ladle.
6. Put in your powder.
7. Empty your ladle.
8. Put up your powder.
9. Thrust home your wad.
10. Regard your shot.
11. Put home your shot gently.
12. Thrust home your wad with three strokes.
13. Gauge your piece.

Gunners had no trouble finding work, as is singularly illustrated by the case of Andrew Ransom, a stray Englishman captured near St. Augustine in the late 1600's. He was condemned to death. The executional device

80

failed, however, and the padres in attendance took it as an act of God and led Ransom to sanctuary at the friary. Meanwhile, the Spanish governor learned this man was an artillerist and a maker of "artificial fires." The governor offered to "protect" him if he would live at the Castillo and put his talents to use. Ransom did.

By 1800, although guns could be served with as few as three men, efficient drill usually called for a much larger force. The smallest crew listed in the United States Navy manual of 1866 was seven: first and second

FIGURE 49—A SIEGE BOMBARD OF THE 1500's.

gun captains, two loaders, two spongers, and a "powder monkey" (powder boy). An 11-inch pivot-gun on its revolving carriage was served by 24 crewmen and a powderman. In the field, transportation for a 24-pounder siege gun took 10 horses and 5 drivers.

Twelve rounds an hour was good practice for heavy guns during the Civil War period, although the figure could be upped to 20 rounds. By this date, of course, although the principles of muzzle loading had not changed, actual loading of the gun was greatly simplified by using fixed and semi-fixed ammunition. Loading technique varied with the gun, but the following summary of drill from the United States *Heavy Ordnance Manual* of 1861 gives a fair idea of how the crew handled a siege gun:

In the first place, consider that the equipment is all in its proper place. The gun is on a two-wheeled siege carriage, and is "in battery," or pushed forward on the platform until the muzzle is in the earthwork embrasure. On each side of the gun are three handspikes, leaning against the parapet. On the right of the gun a sponge and a rammer are laid on a prop, about 6 feet away from the carriage. Near the left muzzle of the gun is a stack of cannonballs, wads, and a "passbox" or powder bucket. Hanging from the cascabel are two pouches: the tube-pouch containing friction "tubes" (primers for the vent) and the lanyard; and the gunner's-pouch with the gunner's level, breech-sight, pick, gimlet, vent-punch, chalk, and finger-stall (a leather cover for the gunner's second left finger when the gun gets hot). Under the wheels are two chocks; the vent-cover is on the vent, a tompion in the muzzle; a broom leans against the parapet beyond the stack of cannonballs. A wormer, ladle, and wrench were also part of the battery equipment.

The crew consisted of a gunner and six cannoneers. At the command *Take implements* the gunner stepped to the cascabel and handed the vent-cover to No. 2; the tube-pouch he gave to No. 3; he put on his fingerstall, leveled the gun with the elevating screw, applied his level to base ring and muzzle to find the highest points of the barrel, and marked these points with chalk for a line of sight. His six crewmen took their positions about a yard apart, three men on each side of the gun, with handspikes ready.

From battery was the first command of the drill. The gunner stepped from behind the gun, while the handspikemen embarred their spikes. Cannoneers Nos. 1, 3, and 5 were on the right side of the gun, and the even-numbered men were on the left. Nos. 1 and 2 put their spikes under the front of the wheels; Nos. 3 and 4 embarred under the carriage cheeks to bear down on the rear spokes of the wheel; Nos. 5 and 6 had their spikes under the maneuvering bolts of the trail for guiding the piece away from the parapet. With the gunner's word *Heave,* the men at the wheels put on the pressure, and with successive *heaves* the gun was moved backward until the muzzle was clear of the embrasure by a yard. The crew then unbarred, and Nos. 1 and 2 chocked the wheels.

Load was the second command. Nos. 1, 2, and 4 laid down their spikes; No. 2 took out the tompion; No. 1 took up the sponge and put its wooly head into the muzzle; No. 2 stepped up to the muzzle and seized the

FIGURE 50—GUN DRILL IN THE 1850's.

sponge staff to help No. 1. In five counts they pushed the sponge to the bottom of the bore. Meanwhile, No. 4 took the passbox and went to the magazine for a cartridge.

The gunner put his finger over the vent, and with his right hand turned the elevating screw to adjust the piece conveniently for loading. No. 3 picked up the rammer.

At the command *Sponge,* the men at the sponge pressed the tool against the bottom of the bore and gave it three turns from right to left, then three turns from left to right. Next the sponge was drawn, and while No. 1 exchanged it for No. 3's rammer, the No. 2 man took the cartridge from No. 4, and put it in the bore. He helped No. 1 push it home with the rammer, while No. 4 went for a ball and, if necessary, a wad.

Ram! The men on the rammer drew it out an arm's length and rammed the cartridge with a single stroke. No. 2 took the ball from No. 4, while No. 1 threw out the rammer. With the ball in the bore, both men again manned the rammer to force the shot home and delivered a final single-stroke ram. No. 1 put the rammer back on its prop. The gunner stuck his pick into the vent to prick open the powder bag.

The command *In battery* was the signal for the cannoneers to man the handspikes again, Nos. 1, 2, 3, and 4 working at the wheels and Nos. 5 and 6 guiding the trail as before. After successive *heaves,* the gunner halted the piece with the wheels touching the hurter—the timber laid at the foot of the parapet to stop the wheels.

Point was the next order. No. 3, the man with the tube-pouch, got out his lanyard and hooked it to a primer. Nos. 5 and 6 put their handspikes under the trail, ready to move the gun right or left. The gunner went to the breech of the gun, removed his pick from the vent, and, sighting down the barrel, directed the spikemen: he would tap the right side of the breech, and No. 5 would heave on his handspike to inch the trail toward the left. A tap on the left side would move No. 6 in the opposite direction. Next, the gunner put the breech-sight (if he needed it) carefully on the chalk line of the base ring and ran the elevating screw to the proper elevation.

As soon as the gun was properly laid, the gunner said *Ready* and signaled with both hands. He took the breech-sight off the gun and walked over to windward, where he could watch the effect of the shot. Nos. 1 and 2 had the chocks, ready to block the wheels at the end of the recoil. No. 3 put the primer in the vent, uncoiled the lanyard and broke a full pace to the rear with his left foot. He stretched the lanyard, holding it in his right hand.

At *Fire!* No. 3 gave a smart pull on the lanyard. The gun fired, the carriage recoiled, and Nos. 1 and 2 chocked the wheels. No. 3 rewound his lanyard, and the gunner, having watched the shot, returned to his post.

The development of heavy ordnance through the ages is a subject with many fascinating ramifications, but this survey has of necessity been brief.

It has only been possible to indicate the general pattern. Most of the interesting details must await the publication of much larger volumes. It is hoped, however, that enough information has been included herein to enhance the enjoyment that comes from inspecting the great variety of cannon and projectiles that are to be seen throughout the National Park System.

Glossary

Most technical phrases are explained in the text and illustrations (see fig. 51). For convenient reference, however, some important words are defined below:

Ballistics—the science dealing with the motion of projectiles.

Barbette carriage—as used here, a traverse carriage on which a gun is mounted to fire over a parapet.

Bomb, bombshell—see projectiles.

Breechblock—a movable piece which closes the breech of a cannon.

Caliber—diameter of the bore; also used to express bore length. A 30-caliber gun has a bore length 30 times the diameter of the bore.

Cartridge—a bag or case holding a complete powder charge for the cannon, and in some instances also containing the projectile.

Casemate carriage—as used here, a traverse carriage in a fort gunroom (casemate). The gun fired through an embrasure or loophole in the wall of the room.

Chamber—the part of the bore which holds the propelling charge, especially when of different diameter than the rest of the bore; in chambered muzzle-loaders, the chamber diameter was smaller than that of the bore.

Elevation—the angle between the axis of a piece and the horizontal plane.

Fuze—a device to ignite the charge of a shell or other projectile.

Grommet—a rope ring used as a wad to hold a cannonball in place in the bore.

Gun—any firearm; in the limited sense, a long cannon with high muzzle velocity and flat trajectory.

Howitzer—a short cannon, intermediate between the gun and mortar.

Lay—to aim a gun.

Limber—a two-wheeled vehicle to which the gun trail is attached for transport.

Mandrel—a metal bar, used as a core around which metal may be forged or otherwise shaped.

Mortar—a very short cannon used for high or curved trajectory firing.

Point-blank—as used here, the point where the projectile, when fired from a level bore, first strikes the horizontal ground in front of the cannon.

Projectiles—*canister* or *case shot*: a can filled with small missiles that scatter after firing from the gun. *Grape shot*: a cluster of small iron balls, which scatter upon firing. *Shell*: explosive missile; a hollow cast-iron ball, filled with gunpowder, with a fuze to produce detonation; a long, hollow projectile, filled with explosive and fitted with a fuze. *Shot*: a solid projectile, non-explosive.

Quoin—a wedge placed under the breech of a gun to fix its elevation.

Range—The horizontal distance from a gun to its target or to the point where the projectile first strikes the ground. *Effective range* is the distance at which effective results may be expected, and is usually not the same as *maximum range,* which means the extreme limit of range.

Rotating band—a band of soft metal, such as copper, which encircles the projectile near its base. By engaging the lands of the spiral rifling in the bore, the band causes rotation of the projectile. Rotating bands for muzzle-loading cannon were expansion rings, and the powder blast expanded the ring into the rifling grooves.

Train—to aim a gun.

Trajectory—curved path taken by a projectile in its flight through the air.

Transom—horizontal beam between the cheeks of a gun carriage.

Traverse carriage—as used here, a stationary gun mount, consisting of a gun carriage on a wheeled platform which can be moved about a pivot for aiming the gun to right or left.

Windage—as used here, the difference between the diameter of the shot and the diameter of the bore.

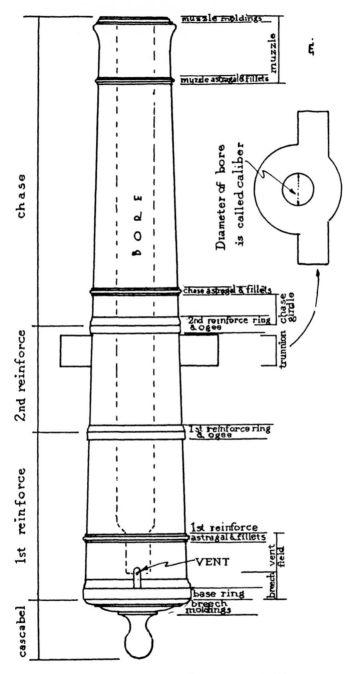

FIGURE 51—THE PARTS OF A CANNON.

Selected Bibliography

The following is a listing of the more important sources dealing with the development of artillery which have been consulted in the production of this booklet. None of the German or Italian sources have been included, since practically no German or Italian guns were used in this country.

SPANISH ORDNANCE. Luis Collado, "Platica Manual de la Altillería" ms., Milan 1592, and Diego Ufano, *Artillerie*, n. p., 1621, have detailed information on sixteenth century guns, and Tomás de Morla, *Láminas pertenecientes al Tratado de Artillería*, Madrid, 1803, illustrates eighteenth century material. Thor Borresen, "Spanish Guns and Carriages, 1686-1800" ms., Yorktown, 1938, summarizes eighteenth century changes in Spanish and French artillery. Information on colonial use of cannon can be found in mss. of the Archivo General de Indias as follows: Inventories of Castillo de San Marcos armament in 1683 (58-2-2,32/2), 1706 (58-1-27,89/2), 1740 (58-1-32), 1763 (86-7-11,19), Zuñiga's report on the 1702 siege of St. Augustine (58-2-8,B3), and Arredondo's "Plan de la Ciudad de Sn. Agustín de la Florida" (87-1-1/2, ms. map); and other works, including [Andres Gonzales de Barcía,] *Ensayo Cronológico para la Historia General de la Florida,* Madrid, 1723; J. T. Connor, editor, *Colonial Records of Spanish Florida,* Deland, 1930, Vol. II., Manuel de Montiano, *Letters of Montiano* (*Collections* of the Georgia Historical Society, v. VII, pt. I), Savannah 1909; Albert Manucy, "Ordnance used at Castillo de San Marcos, 1672-1834," St. Augustine, 1939.

ENGLISH ORDNANCE. For detailed information John Müller, *Treatise of Artillery,* London, 1756, has been the basic source for eighteenth century material. William Bourne, *The Arte of Shooting in Great Ordnance,* London, 1587, discusses sixteenth century artillery; and the anonymous *New Method of Fortification,* London, 1748, contains much seventeenth century information. For colonial artillery data there is John Smith, *The Generall Historie of Virginia, New-Englande, and the Summer Isles,* Richmond, 1819; [Edward Kimber] *Late Expedition to the Gates of St. Augustine,* Boston, 1935; and C. L. Mowat, *East Florida as a British*

Province, 1763-1784, Los Angeles, 1939. Charles J. Ffoulkes, *The Gun-Founders of England*, Cambridge, 1937, discusses the construction of early cannon in England.

FRENCH ORDNANCE. M. Surirey de Saint-Remy, *Memories d'Artillerie*, 3rd edition Paris, 1745, is the standard source for French artillery material in the seventeenth and early eighteenth centuries. Col. Favé, *Ètudes sur le Passé et l'Avenir de L'Artillerie*, Paris, 1863, is a good general history. Louis Figurier, *Armes de Guerre*, Paris, 1870, is also useful.

UNITED STATES ORDNANCE. Of first importance is Louis de Tousard, *American Artillerist's Companion*, 2 vols., Philadelphia, 1809-13. For performance and use of artillery during the 1860's the following sources are useful: John Gibbon, *The Artillerist's Manual*, New York, 1863; Q. A. Gillmore, *Engineer and Artillery Operations against the Defences of Charleston Harbor in 1863*, New York, 1865; his *Official Report . . . of the Siege and Reduction of Fort Pulaski, Georgia*, New York, 1862; and the *Official Records of Union and Confederate Armies and Navies*. Ordnance manuals of the period include: *Instruction for Heavy Artillery*, U. S., Charleston, 1861; *Ordnance Instructions for the United States Navy*, Washington, 1866; J. Gorgas, *The Ordnance Manual for the Use of the Officers of the Confederate States Army*, Richmond, 1863. For United States developments after 1860: L. L. Bruff, *A Text-book of Ordnance and Gunnery*, New York, 1903; F. T. Hines and F. W. Ward, *The Service of Coast Artillery*, New York, 1910; the U. S. Field Artillery School's *Construction of Field Artillery Matériel* and *General Characteristics of Field Artillery Ammunition*, Fort Sill, 1941.

GENERAL. For the history of artillery, as well as additional biographical and technical details, there is the Field Artillery School's excellent booklet, *History of the Development of Field Artillery Matériel*, Fort Sill, 1941. Henry W. L. Hime, *The Origin of Artillery*, New York, 1915, is most useful, as is that standard work, the *Encyclopedia Britannica*, 1894 edition: Arms and Armour, Artillery, Gunmaking, Gunnery, Gunpowder; 1938 edition: Artillery, Coehoorn, Engines of War, Fireworks, Gribeauval, Gun, Gunnery, Gunpowder, Musket, Ordnance, Rocket, Smallarms, and Tartaglia.

Printed in the United States
5540